Computational Intelligence based Optimization of Manufacturing Process for Sustainable Materials

This book comprehensively discusses computational models, including artificial neural networks, agent-based models, and decision field theory for reliability engineering. It will serve as an ideal reference text for graduate students and academic researchers in the fields of industrial engineering, manufacturing engineering, computer engineering, and materials science. In short, the book:

- Discusses the development of sustainable materials using metaheuristic approaches.
- Covers computational models such as agent-based models, ontology, and decision field theory for reliability engineering.
- Presents swarm intelligence methods such as ant colony optimization, particle swarm optimization, and grey wolf optimization for solving the manufacturing process.
- Includes case studies for industrial optimizations.
- Explores the use of computational optimization for reliability and maintainability theory.

The text covers swarm intelligence techniques, including ant colony optimization, particle swarm optimization, cuckoo search, and genetic algorithms for solving the complex industrial problems of manufacturing industry as well as predicting reliability, maintainability, and availability of several industrial components.

Computational and Intelligent Systems Series

In today's world, the systems that integrate intelligence into machine-based applications are known as intelligent systems. In order to simplify man–machine interaction, intelligent systems play an important role. The books under the proposed series will explain the fundamentals of intelligent systems, review the computational techniques and also offer step-by-step solutions of the practical problems. Aimed at senior undergraduate students, graduate students, academic researchers and professionals, the proposed series will focus on broad topics, including artificial intelligence, deep learning, image processing, cyber-physical systems, wireless security, mechatronics, cognitive computing, and Industry 4.0.

Application of Soft Computing Techniques in Mechanical Engineering
Amar Patnaik, Vikas Kukshal, Pankaj Agarwal, Ankush Sharma and Mahavir Choudhary

Computational Intelligence based Optimization of Manufacturing Process for Sustainable Materials
Deepak Sinwar, Kamalakanta Muduli, Vijaypal Singh Dhaka and Vijander Singh

Computational Intelligence based Optimization of Manufacturing Process for Sustainable Materials

Edited by
Deepak Sinwar
Kamalakanta Muduli
Vijaypal Singh Dhaka
Vijander Singh

CRC Press
Taylor & Francis Group
Boca Raton London New York

CRC Press is an imprint of the
Taylor & Francis Group, an **informa** business

First edition published 2024
by CRC Press
6000 Broken Sound Parkway NW, Suite 300, Boca Raton, FL 33487-2742

and by CRC Press

4 Park Square, Milton Park, Abingdon, Oxon, OX14 4RN

CRC Press is an imprint of Taylor & Francis Group, LLC

ISBN: 978-1-032-19104-1 (hbk)
ISBN: 978-1-032-19110-2 (pbk)
ISBN: 978-1-003-25771-4 (ebk)

DOI: 10.1201/9781003257714

Typeset in Sabon
by SPi Technologies India Pvt Ltd (Straive)

Contents

Preface

Computational Intelligence (CI) has been a rapidly growing field in recent years, with increasing research and development efforts aimed at developing and applying advanced computational techniques to solve complex problems in various domains. One such domain that has seen significant progress in the application of CI is manufacturing, specifically the optimization of manufacturing processes for sustainable materials.

This book presents an overview of the state of the art in the use of CI techniques for the optimization of manufacturing processes for sustainable materials. The book is intended for researchers, engineers, and practitioners in the field of manufacturing, materials science, and computational intelligence. The book covers not only the basic concepts and techniques of CI but also various applications of CI in the manufacturing processes for sustainable materials. CI techniques are widely used in the optimization of manufacturing processes and are essential for understanding sustainable manufacturing. The chapters in this book provided detailed descriptions of the methods used, as well as case studies demonstrating the effectiveness of these methods in real-world applications. The chapter-wise summary is provided as follows:

Chapter 1 introduces the role of CI for sustainable materials. The authors have focused on reducing the carbon footprint with the help of CI-based techniques. They also presented several optimization algorithms, along with their applications for sustainable materials.

Chapter 2 presents in-depth research into the areas of the Internet of Things (IoT) and artificial intelligence (AI) to enable the design of sustainable manufacturing for Industry 4.0. They discuss the need and role of AI-enabled sustainable manufacturing. In addition, the role of Industry 4.0 and IoT in different industries is also discussed.

Chapter 3 explores the variety of defects and their classifications identified in manufacturing industries such as foundry casting, casting plates, metal surfaces, welding spots, plane surfaces and barrel surfaces in turbine blades. Secondly, it reviews the machine learning mechanisms, discriminations on various approaches, listing the pros and cons of each course and summarizing the methods used in the existing system and the experimental result achieved.

Chapter 4 focuses on performance optimization and the prediction of manufacturing data. They discussed not only the production planning and control but also its applications and dynamic scheduling algorithms. In addition, the application of artificial intelligence and machine learning talents allows for enhanced adaptability and flexibility, which is essential for achieving high levels of performance.

Chapter 5 discovers concrete's potential by utilizing bentonite, Recycled Glass Aggregate (RGA) and Recycled Concrete Aggregate (RCA). The study deals with the examination of compressive strength, workability and split tensile strength. Based on the test results, the maximum compressive strength, workability and split tensile strength for the optimal percentage of mix was found out using a Multi-Criteria Decision Making (MCDM) technique known as Multi-Objective Optimization based on Ratio Analysis (MOORA).

Chapter 6 emphasizes the recent advancements in High Entropy Alloys (HEA) for thermoelectric applications. The high entropy material design and their impact over various thermodynamic characteristics, thermoelectric properties, viable synthesis routes, strategies to enhance the thermoelectric performance, and their limitations are systematically correlated. Further, this chapter also discusses the role of density functional theory (DFT)-based first principle studies in predicting various underlying principles in thermoelectric performances, such as band structure modification, the electron density of states, thermodynamic parameters of the high entropy alloys, etc.

Chapter 7 presents the incorporation of augmented reality (AR) and virtual reality (VR) in the manufacturing industry 4.0. They provided the fundamentals of AR and VR, conducted a literature review to identify their strengths and limitations, and provide an overview of the research work done by researchers on AR and VR implementation in various sectors. The value of AR in many industries is summarized in this study.

Chapter 8 presents the optimization of process parameters that are required to improve fatigue strength with high density and also parts manufactured by the selective laser melting (SLM) of AlSi10Mg alloy. They obtained the metallurgical keyhole pores, cracks, and overlap as thermal deviations produced at high laser power.

Chapter 9 underlines the role of artificial intelligence in the transformation of apparel industry in the context of Industry 4.0 and 5.0. Effectiveness and value creation were central points of Industry 4.0, whereas Industry 5.0 focuses on sustainability and human-centric approaches. The personalization and collaboration of creative and analytical skills are key needs of both Industry 4.0 and 5.0. Artificial intelligence is seen as a key player in this context.

Chapter 10 presents an automotive manufacturing system based on swarm intelligence that would help machines to adapt to such disruptions and the production flow will not be hampered. Each component of the manufacturing system is controlled by respective 'cognitive agents' according to this swarm technology. It helps in autonomous functioning and the recovery of the system without the intervention of enterprise resource planning or the Manufacturing Execution System. The proposed autonomous system also meets the prerequisites of flexibility, versatility, and dependability.

Overall, this book aims to provide a comprehensive overview of the current state of the art in the use of computing intelligence techniques for the optimization of manufacturing processes for sustainable materials. It is our hope that the book will inspire further research and development in this important and rapidly evolving field.

Editors

Dr. Deepak Sinwar is an Associate Professor at the Department of Computer and Communication Engineering, Manipal University Jaipur, Jaipur, Rajasthan, India. He earned his Ph.D and M.Tech degrees in Computer Science and Engineering in 2016 and 2010, respectively; and his B.Tech (with honors) in Information Technology in 2008. He is an enthusiastic and motivating academician with more than 12 years of teaching experience. His research interests include computational intelligence, data mining, machine learning, reliability theory, computer networks and pattern recognition. On his credit, he has published more than 50 articles in peer reviewed journals, conference proceedings, and book chapters. He has been involved in many editorial activities like editing books/ special issues with publishers of repute like Taylor & Francis, SpringerNature, Wiley, IGI Global, etc. He has organized and attended various conferences and workshops during his teaching career. He is a life member of Indian Society for Technical Education (ISTE), senior member of IEEE and member of ACM professional society.

Dr. Kamalakanta Muduli is an Associate Professor in the Department of Mechanical Engineering, Papua New Guinea University of Technology, Lae, Morobe Province, Papua New Guinea. He is also a visiting researcher at the University of Derby, UK. He earned his PhD from the School of Mechanical Sciences, IIT Bhubaneswar, Orissa, India. Dr. Muduli has more than 16 years of academic experience in universities in India and Papua New Guinea. Dr. Muduli is a recipient of the ERASMUS+ KA107 award provided by the European Union. He has published 69 papers in peer-reviewed international journals, most of which are indexed in Clarivate analytics, Scopus and listed in ABDC. He has also presented more than 34 papers in national and international conferences. He has been also guest-edited 5 special issues of journals and 8 books published by Springer, Taylor & Francis, MDPI, CRC Press, Wiley scrivener and Apple Academic Press. Dr. Muduli has also guided three PhD students. His current research interests include materials science, manufacturing, sustainable supply chain management, and Industry 4.0 applications in operations and supply chain management. Dr. Muduli is a fellow of Institution of Engineers India. He is also a senior member of the Indian Institution of Industrial Engineering and a member of ASME.

Prof. Vijaypal Singh Dhaka is a seasoned academician with an entrepreneurial spirit. He enjoys a passion for continuous learning, both for himself and his students. He has more than 17 years of experience in the software industry, the academic world, and also in research, teaching, and training. His continuous diligence in refining academic quality, building research acumen, developing entrepreneurial culture, improving experiential learning and creating a fostering culture for innovation has facilitated the students of Manipal University Jaipur in achieving laurels at various platforms. In addition, he has more than 100 publications in journals of great repute in his name and has guided 13 research scholars to earn their Ph.Ds. He is also executing DST-sponsored research projects and supervising Ph.D. research scholars and many funded student projects so that strong concepts can lead to great professional skills.

Dr. Vijander Singh earned his Ph.D. degree from Banasthali University, Banasthali, India, in April 2017, and the M.Tech. degree (Hons.) from Rajasthan Technical University, in 2019. He also qualified NET and GATE examinations, in 2012 and 2018, respectively. He is a Postdoctoral Fellow at the Cyber-Physical Systems Laboratory, Department of ICT and Natural Sciences, Faculty of Information Technology and Electrical Engineering, Norwegian University of Science and Technology, Norway, and an Associate Professor with the Department of Computer Science and Engineering, Manipal University Jaipur, Jaipur, India. He has published more than 40 journal articles, 15 conference papers, 10 book chapters, and 2 edited books. His research interests include machine learning, deep learning, precision agriculture, and networking. He is also a guest editor for journals of international repute. He has organized several international conferences, FDPs, and workshops, as a core team member of their organizing committees.

Contributors

A. Anandraj
Saranathan College of Engineering, Panjappur, Tamilnadu, India

Mahin Anup
Microsoft India (R&D) Pvt. Ltd. Hyderabad, Telangana, India

Tulasi B.
CHRIST (Deemed to be University) Bangalore, Karnataka, India

Dilliraj Ekambaram
SRM Institute of Science and Technology
Chennai, Tamil Nadu, India

N. Ethiraj
Dr. M.G.R. Educational and Research Institute
Maduravoyal, Tamil Nadu, India

Kamal Golui
Gargi Memorial Institute of Technology, Baruipur, Kolkata, India

K. Jayanthi
KGiSL Institute of Information Management
Coimbatore, Tamil Nadu, India

A. Joshi
Panimalar Engineering College Chennai, Tamil Nadu, India

N. Kanya
Dr. M.G.R. Educational and Research Institute
Maduravoyal, Tamil Nadu, India

Sudhir Karanam
Reliance Retail, India
Bangalore, Karnataka, India

M. Kiruthiga Devi
Dr. M.G.R Educational and Research Institute, Chennai, Tamil Nadu, India

Amit Mithal
Jaipur Engineering College & Research Centre
Jaipur, Rajasthan, India

Rohit Mittal
Manipal University Jaipur Jaipur, Rajasthan, India

Mudda Nirish
Osmania University Hyderabad, Telangana, India

P. Vivekanandhan
Centre for Automotive Energy
  Technology, International
  Advanced Research Centre for
  Powder Metallurgy and New
  Materials, Chennai, Tamil Nadu,
  India
and
National Institute of Technology,
  Tiruchirappalli, Tamil Nadu, India

Vibhakar Pathak
Arya College of Engineering & IT
Jaipur, Rajasthan, India

Vijayakumar Ponnusamy
SRM Institute of Science and
  Technology
Chennai, Tamil Nadu, India

R. Rajendra
Osmania University,
Hyderabad, Telangana, India

P.V. Rajesh
Saranathan College of Engineering,
  Panjappur, Tamilnadu, India

Arun Raphel
National Institute of Technology,
  Tiruchirappalli, Tamil Nadu, India
and
Viswajyothi College of Engineering
  and Technology, Ernakulam, India

S. Kumaran
National Institute of Technology,
  Tiruchirappalli, Tamil Nadu, India

Dahlia Sam
Dr. M.G.R. Educational and Research
  Institute
Maduravoyal, Tamil Nadu, India

Hiranmoy Samanta
Gargi Memorial Institute of
  Technology, Baruipur, Kolkata,
  India

Devika Sapra
Manipal University Jaipur
Jaipur, Rajasthan, India

S. Sendilvelan
Dr. M.G.R. Educational and Research
  Institute
Maduravoyal, Tamil Nadu, India

Akruti Sinha
North Carolina State University
Raleigh, NC, USA

Deepak Sinwar
Manipal University Jaipur
Jaipur, Rajasthan, India

Gaurav Srivastava
Manipal University Jaipur
Jaipur, Rajasthan, India

S. Vijayabaskaran
Saranathan College of Engineering,
  Panjappur, Tamilnadu, India

Chapter 1

Introduction to computational intelligence for sustainable materials

Rohit Mittal
Manipal University Jaipur, Jaipur, Rajasthan, India

Vibhakar Pathak
Arya College of Engineering & IT, Jaipur, Rajasthan, India

Amit Mithal
Jaipur Engineering College & Research Centre, Jaipur, Rajasthan, India

CONTENTS

1.1 COMPUTATIONAL INTELLIGENCE

Computational intelligence (CI) is now having an influence on a wider and wider number of industries. For instance, CI is anticipated to have both immediate and long-term effects on the world's productivity, equality and inclusion, environmental results, and a number of other sectors. Both favourable and unfavourable effects on environmental sustainability have been

DOI: 10.1201/9781003257714-1

identified by the reported potential effects of CI. Here, we exhibit and talk about the effects of CI on sustainable materials. Although there is no universally accepted definition of CI, for the purposes of this study, we defined CI as any software technology with at least one of the following capabilities: perception, such as voice, graphic, text-based, and interactive (e.g., face recognition), involving decision-making, prediction from data or interactive computing (e.g., theory development from premises). This perspective covers many different subfields, and, in particular, machine learning. We require a great deal more detail regarding the impact any business is making as sustainability becomes increasingly crucial. We must monitor and report on internal corporate activity that affects ecological responsibility. CI has the capacity to generate huge and considerable carbon emissions. In addition, CI has the capacity to lessen or neutralize those carbon emissions. This chapter demonstrates the CI techniques in sustainable development to determine the growth of any enterprise knowledge niche overlap inside a company based on the definitions of related terms, such as the innovation ecosystem, innovation capacity, and the knowledge niche. This chapter explains the CI techniques in businesses' green innovation system's integration technique for the creation of distinct knowledge niches. Second, this chapter thoroughly and methodically examines the crucial elements influencing an organization's capacity for innovation from the perspectives of enterprise knowledge niche integration competence and green innovation system features. The emphasis and optimization ideas of the firm's policies at each stage of development are offered, which has considerable practical value, based on the model in combination with the viewpoint of the enterprise life cycle, from the standpoint of dynamics and evolution [1, 2].

1.2 A REDUCTION IN THE CARBON FOOTPRINT USING CI TECHNIQUES

Companies should consider collaborating with any cloud provider who is dedicated to lowering their carbon footprint in order to lower their own. It may be preferable for a corporation to outsource its CI training and processing to a data centre cloud provider that can accomplish it rather than developing large internal projects to lessen environmental effects. In several industries, CI may have a net beneficial impact on environmental sustainability. A number of instances are listed below:

1. By more effectively regulating agricultural yields with environmental circumstances, CI can revolutionize productivity in agriculture. While increasing agricultural yields, CI can assist in the minimization of water and fertilizer use.
2. In order to control the demand and supply of renewable energy, CI can leverage intelligent grid systems and strong predictive capabilities.

CI may increase productivity by more precisely anticipating weather patterns, lowering expenses and reducing the production of unneeded carbon emissions.

3. CI can assist in reducing or eliminating waste in water resource management while cutting costs and reducing environmental impact. Utilizing less water will be made possible by CI-driven localized weather predictions.

1.3 COMPUTATIONAL SUSTAINABILITY RESEARCH

We provide examples of computational sustainability research, which has mostly focused on addressing three broad sustainability themes: achieving a balance between environmental and socioeconomic demands; conserving biodiversity; and using renewable and sustainable energy and materials. The subway lines in Figure 1.1 show how this part is structured according to sustainability topics, highlighting crosscutting computational issues.

Development that satisfies current demands while maintaining the capacity for future generations to use computational intelligence algorithms to satisfy their wants. The development of computational sustainability that meets present needs while still preserving the ability for future generations to apply computational intelligence algorithms to satiate their own desires.

A roadmap for ensuring a successful and more sustainable future for society is set forth in the 2030 UN agenda for sustainable development. A key component of this plan is the development of new materials and chemicals

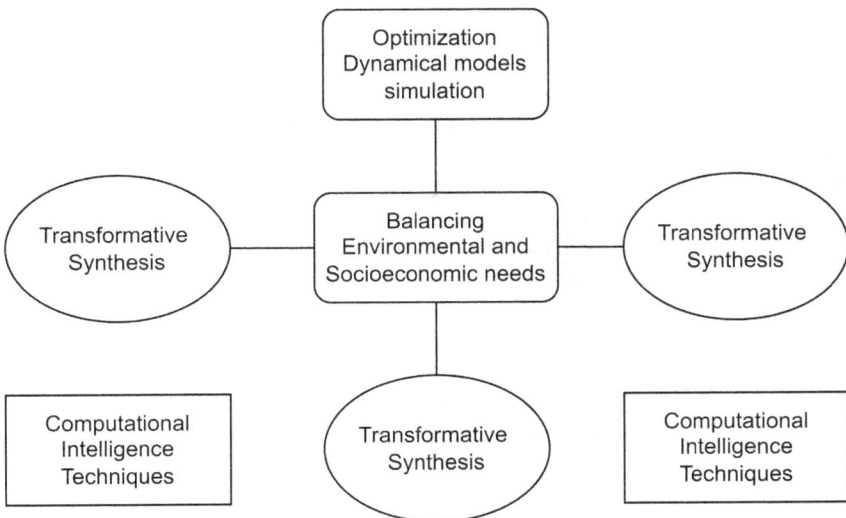

Figure 1.1 The sustainability domains with computational intelligence techniques.

with a focus on sustainability. Conventional materials discovery involves changing one variable at a time, an approach which has several restrictions and disadvantages. Among other disadvantages, experimental datapoints have intrinsic constraints, potential factor interactions are not exposed, and the optimal is rarely attained. The design of experiments (DoE) is a different systematic technique that successfully strikes a balance between efficiency and fewer experimentation. DoE enables the simultaneous investigation and variation of the elements, speeding up the discovery and optimization process while saving valuable resources, labour, and time, eventually leading to a more sustainable strategy [2–4]. There is a need to create and evaluate approaches that can translate the limited experimental results into considerably larger virtual datasets since DoE only uses a small amount of experimental data.

1.4 TECHNIQUES FOR COMPUTATIONAL INTELLIGENCE OPTIMIZATION

Recursive optimization has grown in prominence during the last two decades. Unexpectedly, many of them, such as biological algorithms like Genetic Algorithm (GA), Ant Colony Optimization (ACO), and Particle Swarm Optimization (PSO) [3], are renowned among researchers and other scientists. These optimization methods have been applied in many theoretical papers and academic domains.

The four essential factors in the response to this popularity are simplicity, adaptability, a derivation-free method, and the prevention of local optimum. Most of these algorithms have been influenced by relatively straightforward ideas. Nature often inspired events, such as animal behaviour or adaptive theories. The ease with which computer scientists may duplicate many natural conceptions, propose new meta-heuristics, integrate two or more meta-heuristics, or improve upon existing meta-heuristics is due in large part to the simplicity of these tools. Furthermore, the simplicity makes it easier for other researchers to pick up quickly on meta-heuristics and use them to solve their issues. Second, flexibility is the capacity of meta-heuristics to be applied to many problems without requiring particular alterations to the algorithm's structure. Since meta-heuristics typically treat issues as "black boxes," they can be used to solve a variety of difficulties. In other words, recursive optimization algorithms consider only the inputs and outputs of a system. Thus, all a designer has to know is how to model their issue for recursiveness. Third, the bulk of recursive optimization use processes without derivation. The techniques stochastically optimize issues as opposed to gradient-based optimization techniques. There is no need to compute the derivative of search spaces to identify the optimum because the optimization process starts with random solution(s). Swarm Intelligence (SI) offers a fascinating subset of population-based recursive optimization techniques (SI) [4].

SI is "The emergent collective intelligence of groupings of simple agents," according to Bonabeau et al. [5]. The primary sources of inspiration for SI approaches are natural flocks, herds, colonies, and schools. ACO [6], PSO [7], and Artificial Bee Colony (ABC) [8] are among the SI approaches that are used most often. SI algorithms have the following benefits over Evolutionary Methods (EM), which discard knowledge from previous generations:

- SI algorithms hold data about the problem space during the course of iteration.
- SI algorithms are simple to build and typically have fewer parameters to change than evolutionary techniques (crossover, mutation, elitism, and so on).
- SI algorithms also frequently use memory to preserve the best solution found to date.

1.4.1 Grey wolf optimization

The grey wolf, which hunts enormous prey in groups and depends on inter-pack cooperation, provided the idea for this algorithm. This behaviour has two intriguing aspects:

1. Social standing
2. Mechanism for hunting

The grey wolf has a complicated social structure as a result of being a highly gregarious species. The term "dominance hierarchies" refers to this system of ranking wolves according to their size and power. There were also the alphas, betas, deltas, and omegas as a result.

The alpha males and females are the most dominant members of the pack, and they take the role of leader. Every member of the pack is then ranked according to their position. The wolf pack's hierarchy helps weaker members of the pack who are unable to hunt for themselves; in other words, it is not simply about hostility and power [4].

The beta wolf comes next. They help the alpha wolf in making choices and maintaining order in the pack.

The gamma wolf is ranked behind the beta wolf. They are frequently powerful but lack the leadership abilities or self-assurance to assume leadership roles.

Last but not least, the omega wolf has no power at all, and other wolves will run after him right away. The omega wolf is also in charge of keeping an eye on the young wolves.

In addition to maintain a distinct social structure, grey wolves have a highly particular hunting style. They hunt in packs and cooperate to separate the prey from the herd. One or two wolves will then pursue and attack the victim, and the rest will chase off any stragglers. According to Muro et al. [9], the wolf pack's hunting tactics include:

1. Pursue after, approach, and track the prey
2. The target is pursued, harassed, and surrounded until it stops moving.
3. When the prey is worn out, assault it.

Mucherino and Seref [10] have examined the scientific uses of GWO. They claimed that GWO has demonstrated promising outcomes in several optimization issues. The outstanding features of this method above other swarm intelligence techniques may account for the high degree of GWO success in solving optimization problems in the literature. This review emphasizes that GWO needs only a small number of parameters and requires no knowledge of the search space's derivation. In addition, GWO is a clear, simple, versatile, and scalable system. The search procedure for the algorithm benefits from a balance between exploration and exploitation, producing a good convergence. As a result, researchers in a variety of scientific and technical sectors have expressed interest in the GWO [11]. Yang and Deb [12] have examined the use of GWO in computational fluid dynamic experiments. Numerous engineering applications have made use of the GWO and its modified versions; these include attribute selection classification methods [13, 14], path planning [15, 16] groundwater remediation and the prediction of soil electrical conductivity, a groundwater remediation method [17, 18].

1.4.1.1 Applications of GWO in sustainability

The GWO algorithm was used to optimize the building's energy use. The Building Energy Optimization (BEO) challenge consists of three parts. Building energy modelling is a method for simulating buildings and calculating energy consumption. The second component, the optimization strategy, modifies the construction parameters to get the desired result by using the simulation results. The relationship between the optimization technique and simulations is the third element [19].

1.4.2 Ant Colony Optimization (ACO)

Concerns about optimal performance are of the utmost importance in the industrial and scientific arenas. In the actual world, examples of these optimization tasks include the following: the group-shop scheduling problem; timetable scheduling; capacity planning; difficulties associated with travelling salesmen; challenges associated with vehicle routing; and portfolio optimization, among other examples. Because of this, a number of different optimization strategies are developed. Among them is the improvement of ant colonies' overall efficiency. Ant colony optimization, often known as ACO, is a probabilistic strategy that may be used to determine which paths are the most effective. To solve a wide range of computing problems, researchers and academics in the field of computing often turn to the ACO approach. In the 1990s, Marco Dorigo was the one who first suggested ACO. This method was devised to describe the process by which ants forage in order to locate a path between their colony and a source of food. The well-known problem of what to do with touring salespeople was originally the impetus for its use. In later years, it was used to the resolution of a wide range of difficult optimization problems [2].

Ants are examples of social insects since they live in colonies. The search for food is the ant's fundamental objective, and it is this goal that directs its conduct. Ants may be seen checking about their nests while scurrying around in their nests. During its quest for food, an ant will often jump back and forth between two locations. An organic substance that has characteristics similar to those of a pheromone is left behind by it as it moves over the ground as it moves. Pheromone trails, which ants leave behind, allow for inter-ant communication. Ants are able to communicate with one another. When an ant finds food, it immediately consumes as much as it can get its mouths on. When it returns, it disperses pheromones along the pathways, and the amount and quality of the food determines which pheromones it leaves behind. Once they get a whiff of that path, more ants will follow in its footsteps. There is a correlation between the quantity of pheromone present and the chance of using that way; the more ants that travel that route, the greater the concentration of pheromone [20].

1.4.2.1 Application of ACO in sustainability

In addition to being a significant contributor to regional and national economic growth, businesses also serve as the incubator for new ideas. Some of these businesses are susceptible to the issue of unsustainable innovation and insufficient dynamic adaptation. They must differentiate themselves from other nations in order to overcome the technological barrier which they are now confronting. This article, which is based on the idea of innovation ecology, carries out extensive study on how to enhance businesses' capacity for green innovation, gathers some important findings, and offers some policy recommendations for the growth of businesses [2].

1.4.3 Applications of the Monarch Butterfly Optimization method in sustainability

Two equal-sized subpopulations in MBO go by the names of subpopulation1 and suppopulation2. Bear in mind that subpopulation1 is made up of those inhabitants with the greatest fitness values, whereas subpopulation2 is made up of the remaining inhabitants. The migration operator and the butterfly regulating operator are therefore the two techniques employed in MBO. The two subpopulations are once again grouped into a population when the global ideal information is preserved after one repetition. The entire population is then divided into two subpopulations in accordance with the updated fitness value. Up until the termination condition is satisfied, the process is repeated [21].

According to recent research, buildings account for 36% of the world's CO_2 emissions and 40% of the world's energy consumption, respectively [22]. The use of lighting and air conditioning accounts for around 57% of the energy used by buildings [23]. Therefore, it is obvious that lowering the energy requirements of buildings is a crucial job [24, 25]. Building optimization is a useful strategy that may progressively reduce a building design's energy consumption. However, a structure's energy requirements depend on its location, use pattern, climate, and building materials. Each design parameter's adjustment might have non-linear effects on the other parameters. As a result, the problem of construction optimization is both difficult and nonlinear. The design specifications of a building, which are frequently modifiable, have an impact on how much energy it uses. For instance, it would be simple to change the outside wall thickness, window size, hangings, and other specifications during the design phase. The examination of the literature shows that changing the design characteristics may significantly lower the energy consumption of a building [26]. In order to minimize building energy consumptions. Academicians have recently focused on building optimization problems (BOP) [27]. Building models have also been created using CI methods [28]. A few building design optimization strategies were examined by Li et al. [29]. BOP has been studied using the techniques of PSO and GA. These methods can successfully avoid local optimums and do not require function gradients. These cutting-edge techniques employ systematic search tactics, necessitating several build simulations and having a sluggish convergence rate. Some of the BOP may take months since each building simulation is a time-consuming phase with a large computational cost [30]. Additionally, because the BOP are non-linear issues, several optimization techniques may become stuck in a local extremum and fail to locate the overall optimum [31]. Therefore, in order to cope with BOP, new solution methods with new capabilities are widely desired. To reduce the energy consumption of a building design, Attia et al. [32] used GA, sequential quadratic programming, and simulated annealing techniques. In addition, a trade-off between the price of a media centre in Paris and the lighting performance was sought using the ACO approach [33]. The goal was to identify the ideal set of GA optimized parameters for BOP [34]. A neural network was used by Shea et al. [35] to understand how an office building uses energy. The best building control settings were then discovered using GA. Using an optimization method, the office's energy usage might be cut by as much as 35%. To reduce building energy usage, some used an unique distributed reinforcement learning technique (a sort of machine learning technology). Compared to several common optimization techniques, the approach could optimize the building more effectively. Many recent research have also incorporated machine learning methodologies. There is no universal optimization strategy that might outperform all of the optimization methods described in the literature since each optimization method has certain benefits and drawbacks. For a certain class of optimization problems, some optimization techniques may perform better. BOP was subjected to several optimization techniques by [36, 37]. The authors discovered that Nelder Mead Simplex (NMS) might effectively reduce an office building's energy usage. The size of the air conditioning units and building envelopes were also optimized using the GA and PSO [38, 39].

1.4.4 Application of Harris Hawks Optimization (HHO) in sustainability

The notable swarm-based optimization method Harris Hawks Optimization (HHO) characterized by a large number of ongoing and time-varying phases of exploration and exploitation. Due to its adaptable structure, superior performance, and superior outcomes, this algorithm has attracted significant interest from scholars since it was originally published in the esteemed *Journal of Future Generation Computer Systems* (FGCS) in 2019. The primary rationale for the HHO approach is based on the coordinated actions and "surprise pounce" pursuit techniques of Harris Hawks in the wild.

The dynamic randomized time-varying character of the escaping energy parameter might further enhance and harmonize the exploitative and exploring behaviours of HHO. This element helps HHO in executing a seamless transition between exploration and extraction. Utilizing a variety of exploration tactics in regard to the conventional positioning of hawks is one way that the exploring tendencies of HHO may be boosted during the first rounds of the process.

In order to increase the superiority of ideas and the intensifying abilities of HHO throughout the optimization process, the progressive selection technique allows search agents to progressively improve their position so that they may only pick a job that is of a higher quality. Before selecting the best possible next step, HHO performs a variety of different search strategies. Additionally, this trait positively affects HHO's propensity for exploitation. Candidate solutions may benefit from the randomized jump strength in balancing their exploration and exploitation inclinations.

1.4.4.1 Optimal power flow taking environmental emissions

One of the most important methods utilized in energy management systems for decades to date for the dependable operation and appropriate planning of contemporary power systems is optimal power flow (OPF). This non-linear, non-convex, multi-dimensional optimization problem has discrete control variables transformer tap ratios and shunt capacitor and continuous variables voltage magnitude and actual power production [40, 41]. In order to run the system effectively and inexpensively in response to the ongoing change in load demand, these factors are modified. Thus, the purpose of OPF is to maximize a few specific system goals including fuel cost, real and reactive power loss, voltage stability improvement, and environmental emissions while maintaining equality and inequality limitations [42]. By applying the HHO method to solve single- and multi-objective OPFs, the fuel cost, power loss, and pollution cost of the system are optimized [43, 44].

1.5 DISCUSSION AND CONCLUSION

This chapter discusses the fundamentals of CI-based optimization techniques such as Grey Wolf Optimization, Ant Colony Optimization, Monarch Butterfly optimization and Harris Hawks Optimization and applications in sustainability. The purpose of this chapter is to serve as a helpful one-stop informational resource for scholars and professionals working in the aforementioned disciplines of study. The study of the enterprise green innovation system and its development in this chapter directly addresses the current issue with enterprise construction. The building's energy usage was optimized using the GWO, ACO Monarch Butterfly optimization and the Harris Hawks Optimization algorithm. The components to the Building Energy Optimization problem can be progressively reduced by building design's energy consumption and optimizing the power flow in these types of building.

However, several of the optimized methods, such as the Elephant Herding Optimization, Differential Evolution, and Covariance Matrix Adaptation Evolution Strategy, are still understudied in sustainable materials. The effectiveness of these new techniques has not, however, been examined for BOP. Therefore, there is a great need to investigate innovative BOP optimization techniques. The innovative BOA can be employed as a method for resolving issues with global optimization. The technique imitates how butterflies find food and mate. Butterflies can utilize their sense of smell to find a mate or a source of nectar. The Building Optimization Approach (BOA) [19] is an appealing approach since it has just a few setting parameters and does not need gradients of the objective function. During the search stages, the algorithm strikes a reasonable balance between exploitation and exploration. The BOA has also been used to identify fuel cell, photovoltaic, and feature selection characteristics. As a result, the BOA was designed to mimic butterflies' keen sense of smell and cooperative behaviour when foraging. The butterflies use this algorithm to release a smell that allows them to communicate with other butterflies. Although BOA was been introduced just recently, it has already been used to optimize a large number of scientific and technical issues.

REFERENCES

[1] T.A. Jumani et al., "Computational Intelligence-Based Optimization Methods for Power Quality and Dynamic Response Enhancement of ac Microgrids," *Energies*, vol. 13, no. 16, p. 4063, 2020, doi:10.3390/en13164063.

[2] Hailin Yang et al., "Integration of Green Innovation Capabilities of Enterprises Based on Ant Colony Optimization Algorithm," *Computational Intelligence and Neuroscience*, vol. 2022, p. 8428641, 2022, doi:10.1155/2022/8428641.

[3] J.H. Holland, "Genetic Algorithms," *Scientific American*, vol. 267, pp. 66–72, 1992.

[4] Seyedali Mirjalili, Seyed Mohammad Mirjalili, and Andrew Lewis, "Grey Wolf Optimizer," *Advances in Engineering Software*, vol. 69, pp. 46–61, 2014, doi:10.1016/j.advengsoft.2013.12.007.

[5] E. Bonabeau, M. Dorigo, and G. Theraulaz, *Swarm Intelligence: From Natural to Artificial Systems*, USA, OUP, 1999.

[6] M. Dorigo, M. Birattari, and T. Stutzle, "Ant Colony Optimization," *IEEE Computational Intelligence Magazine*, vol. 1, pp. 28–39, 2006.

[7] J. Kennedy and R. Eberhart, "Particle Swarm Optimization," in *Proceedings of ICNN'95-International Conference on Neural Networks*, 1995, pp. 1942–1948.

[8] B. Basturk and D. Karaboga, "An Artificial Bee Colony (ABC) Algorithm for Numeric Function Optimization," in *IEEE Swarm Intelligence Symposium*, 2006, pp. 12–14.

[9] C. Muro, R. Escobedo, L. Spector, and R. Coppinger, "Wolf-Pack (Canis Lupus) Hunting Strategies emerge from Simple Rules in Computational Simulations," *Behavioural Processes*, vol. 88, pp. 192–197, 2011.

[10] A. Mucherino and O. Seref, "Monkey Search: A Novel Metaheuristic Search for Global Optimization," in *AIP Conference Proceedings*, 2007, p. 162.

[11] X. Lu and Y. Zhou, "A Novel Global Convergence Algorithm: Bee Collecting Pollen Algorithm," in *Advanced Intelligent Computing Theories and Applications. With Aspects of Artificial Intelligence*, ed: Springer, Berlin, Heidelberg, 2008, pp. 518–525.

[12] X.-S. Yang and S. Deb, "Cuckoo Search via Lévy Flights," in *2009 World Congress on Nature & Biologically Inspired Computing (NaBIC)*, 2009, pp. 210–214.

[13] Y. Shiqin, J. Jianjun, and Y. Guangxing, "A Dolphin Partner Optimization," in *2009 WRI Global Congress on Intelligent Systems*, 2009, pp. 124–128.

[14] X.-S. Yang, "Firefly Algorithm, Stochastic Test Functions and Design Optimisation," *International Journal of Bio-Inspired Computation*, vol. 2, pp. 78–84, 2010.

[15] A. Askarzadeh and A. Rezazadeh, "A New Heuristic Optimization Algorithm for Modeling of Proton Exchange Membrane Fuel Cell: Bird Mating Optimizer," *International Journal of Energy Research*, vol. 37, pp. 1196–1204, 2012.

[16] A.H. Gandomi and A.H. Alavi, "Krill Herd: A New Bio-Inspired Optimization Algorithm," *Communications in Nonlinear Science and Numerical Simulation*, vol. 17, pp. 4831–4845, 2012.

[17] W.-T. Pan, "A New Fruit Fly Optimization Algorithm: Taking the Financial Distress Model as an Example," *Knowledge-Based Systems*, vol. 26, pp. 69–74, 2012.

[18] L.D. Mech, "Alpha Status, Dominance, and Division of Labor in Wolf Packs," *Canadian Journal of Zoology*, vol. 77, pp. 1196–1203, 1999.

[19] Mehdi Ghalambaz, Reza Jalilzadeh, and Amir Davami, "Building Energy Optimization Using Grey Wolf Optimizer (GWO)," *Case Studies in Thermal Engineering*, vol. 27, p. 101250, 2021, doi:10.1016/j.csite.2021.101250.

[20] M. Dorigo, M. Birattari, and T. Stutzle, "Ant Colony Optimization," *IEEE Computational Intelligence Magazine*, vol. 1, no. 4, pp. 28–39, 2006.

[21] G.G. Wang, S. Deb, and Z. Cui, "Monarch Butterfly Optimization," *Neural Computing and Applications*, vol. 31, no. 7, pp. 1995–2014, 2019.

[22] U. Sbci, *Buildings and Climate Change: Summary for Decision-Makers*, United Nations Environ Programme, Sustainable Buildings and Climate Initiative, LOPU, Paris, France, 2009, pp. 1–62.

[23] US Energy Information Administration, *Annual Energy Outlook 2015: With Projections to 2040*, Government Printing Office, Washington DC, USA, 2015.

[24] IPCC, *Climate Change 2007: Synthesis Report*, IPCC, Geneva, 2007.

[25] N. Delgarm, et al., "Multi-Objective Optimization of the Building Energy Performance: A Simulation-Based Approach by Means of Particle Swarm Optimization (PSO)," *Applied Energy*, vol. 170, pp. 293–303, 2016.

[26] T. Li, et al., "Genetic Algorithm for Building Optimization: State-of-the-Art Survey," in *Proceedings, 9th International Conference on Machine Learning and Computing*, Singapore, 2017, pp. 205–210.

[27] R. Guo, et al., "Optimization of Night Ventilation Performance in Office Buildings in a Cold Climate," *Energy and Buildings*, vol. 225, p. 110319, 2020.

[28] M. Wang, et al., "Optimisation of the Double Skin Facade in Hot and Humid Climates through Altering the Design Parameter Combinations," *Building Simulation*, vol. 14, pp. 511–521, 2020.

[29] Z. Li, et al., "A Review of Operational Energy Consumption Calculation Method for Urban Buildings," *Building Simulation*, vol. 13, pp. 739–751, 2020.

[30] A. Li, et al., "Development of an ANN-Based Building Energy Model for Information-Poor Buildings Using Transfer Learning," *Building Simulation*, vol. 14, pp. 89–101, 2020.

[31] V. Machairas, et al., "Algorithms for Optimization of Building Design: A Review," *Renewable and Sustainable Energy Reviews*, vol. 31, pp. 101–112, 2014.

[32] S. Attia, et al., "Assessing Gaps and Needs for Integrating Building Performance Optimization Tools in Net Zero Energy Buildings Design," *Energy and Buildings*, vol. 60, pp. 110–124, 2013.

[33] A.-T. Nguyen, et al., "A Review on Simulation-Based Optimization Methods Applied to Building Performance Analysis," *Applied Energy*, vol. 113, pp. 1043–1058, 2014.

[34] J. Michalek, et al., "Architectural Lay-out Design Optimization," *Engineering Optimization*, vol. 34, no. 5, pp. 461–484, 2002.

[35] K. Shea, et al., "Multicriteria Optimization of Paneled Building Envelopes Using ant Colony Optimization," *Proceedings, Workshop of the European Group for Intelligent Computing in Engineering*, Ascona, Switzerland, 2006, pp. 627–636.

[36] A. Alajmi and J. Wright, "Selecting the Most Efficient Genetic Algorithm Sets in Solving Unconstrained Building Optimization Problem," *International Journal of Sustainable Built Environment*, vol. 3, no. 1, pp. 18–26, 2014.

[37] M. Ilbeigi, et al., "Prediction and Optimization of Energy Consumption in an Office Building Using Artificial Neural Network and a Genetic Algorithm," *Sustainable Cities and Society*, vol. 61, p. 102325, 2020.

[38] Y. Qin, et al., "Energy Optimization for Regional Buildings Based on Distributed Reinforcement Learning," *Sustainable Cities and Society*, vol. 78, p. 103625, 2022.

[39] J. Arroyo, et al., "Reinforced Model Predictive Control (RL-MPC) for Building Energy Management," *Applied Energy*, vol. 309, p. 118346, 2022.

[40] H.W. Dommel and W.F. Tinney, "Optimal Power Flow Solutions," *IEEE Transactions on Power Apparatus and Systems*, vol. PAS-87, pp. 1866–1876, 1968.

[41] K. Abaci and V. Yamacli, "Differential Search Algorithm for Solving Multi-objective Optimal Power Flow Problem," *International Journal of Electrical Power & Energy Systems*, vol. 79, pp. 1–10, 2016.

[42] J.Z. Zhu, "Improved Interior Point Method for OPF Problems—Power Systems," *IEEE Transactions on Power Apparatus and Systems*, vol. 14, pp. 1114–1120, 1999.

[43] V. Veerapandiyan and D. Mary, "Transmission System Reconfiguration to Reduce Losses and Cost Ensuring Voltage Security," *Journal of Power and Energy Engineering*, vol. 4, pp. 4–12, 2016.

[44] M.Z. Islam, et al., "A Harris Hawks Optimization Based Single- and Multi-Objective Optimal Power Flow Considering Environmental Emission," *Sustainability*, vol. 12, no. 13, p. 5248, 2020, doi:10.3390/su12135248.

Chapter 2

Artificial intelligence and IoT-assisted sustainable manufacturing for Industry 4.0

Gaurav Srivastava and Devika Sapra
Manipal University Jaipur, Jaipur, Rajasthan, India

Akruti Sinha
North Carolina State University, Raleigh, NC, USA

Mahin Anup
Microsoft India (R&D) Pvt. Ltd., Hyderabad, Telangana, India

Deepak Sinwar
Manipal University Jaipur, Jaipur, Rajasthan, India

CONTENTS

DOI: 10.1201/9781003257714-2

2.1 INTRODUCTION

An industry refers to a part of an economy that loosely produces or manufactures either raw or finished products or provides services. Since the dawn of civilization, humanity has relied on some degree of industrialization to prosper. The advance of a society is indirectly linked to innovative and ingenious changes in its industries. These technological leaps, which led to fundamental shifts, are today described as "Industrial Revolutions" [1]. The 1st Industrial Revolution was primarily focused in the field of mechanization, the 2nd Industrial Revolution revolved around the intensive use of electrical energy and the 3rd Industrial Revolution is characterized by widespread digitization [2]. This provides the background to the current revolution, Industry 4.0, which was traditionally aimed at digitizing operations and gaining benefits. Fueled by the Internet of Things (IoT), artificial intelligence (AI), machine learning (ML), computer vision, and data analysis, it enables development to be achieved at much lower costs by using energy and resources efficiently. The usage of interlinked technologies powered by IoT enables Industry 4.0 to be highly cost-effective and free from the traditional manufacturing errors generally caused due to human intervention. Applying the basics of IoT-cloud architecture by using sensors and actuators and processing real-time data will help the whole manufacturing process to come under a single roof, thereby enabling a data-driven approach for manufacturing which will allow future industries to reduce the difficulties experienced during decision-making and facilitate the timely exchange of goods and services in the global supply chain. Combining subfields of AI, such as ML, natural language processing (NLP), and computer vision can help the industry develop more efficient processes and thereby reduce energy use. AI, combined with edge computing methods where the data is stored, processed, and managed directly at IoT endpoints, allows the entire manufacturing life cycle to be streamlined. Further, the application of Industry 4.0 enables an energy-efficient and environmentally friendly approach and helps mitigate the carbon footprint of large-scale industries. The end product of creating sustainable manufacturing can be characterized as being resource-efficient and adaptable to the growing demands of consumers and business partners alike. It delivers the ability to respond flexibly to global industrial failures and disruptions such as the recent CCOVID-19 epidemic crippling the supply chain and causing labor shortages and an unsustainable environment for manufacturing.

Continuing the progress made through the 3rd Industrial Revolution and the usage of advanced digitization has resulted in a new paradigm shift in manufacturing. The combination of Internet technologies and future "smart" technologies results in an entirely new branch of industrialization. Following on from the 3rd stage of the Industrial Revolution, developed countries are now moving towards the fourth Industrial Revolution or, as coined by Klaus Schwab, 4IR or Industry 4.0 [3]. Broadly, this can be classified as the amalgamation of IoT, cloud computing, AI, ML, and data analysis. Current and future manufacturing imagines an efficient and streamlined manufacturing system backed by smart technology. This smart technology entails the integration of the physical setting of the factory with IoT and cloud computing. This unique implementation can thus be called Application pull or Technology push [4]. The integrated platform helps to monitor and predict the parameters and operations in real time. The use of smart machines can analyze and diagnose issues without any need for human intervention.

Industry 4.0 can be distinguished as comprising four key components: cyber-physical systems (CPSs), the Internet of Things (IoT), cloud computing, and cognitive computing [5]. Major components of Industry 4.0 are depicted in Figure 2.1. To turn a physical setting such as a factory site into a "smart factory" all four components need to be integrated, thereby enabling a smart environment for streamlined manufacturing. In a CPS, the physical and digital levels merge. This integration is governed by computer-based algorithms. It enables the user and physical system to carry out homogeneous work and to achieve the desired output. The real-time condition of the object arises from its physical condition which, in turn, provides datasets

Internet of Things   **Big Data & Analytics**   **Cloud Computing**

Simulation   **INDUSTRY 4.0**   **Cyber Security**

Additive Manufacturing   **Augmented Reality**   **Automation**

Figure 2.1 Components of Industry 4.0.

that are recorded digitally. This high level of integration can be useful for high productivity. IoT involves the connection of various devices to the Internet and exchanging data. It allows users to connect and interact with various objects and to monitor and control them remotely [6]. Cloud computing refers to a shared pool of available computer resources that can rapidly be used to manage and compute large datasets. It can also be deployed to achieve economic results for the manufacturing site similar to a public utility. Cognitive computing uses AI and ML to better predict accurate models of human response to a stimulus.

2.2 RELATED WORKS

The intelligent manufacturing structure, as defined in Jena et al. [7], is the vertical integration of various components such as industrial networks, clouds, and supervisory control terminals that perform various functions, including production, maintenance, energy consumption, water consumption, and so on. Factories optimize resource utilization and eliminate all sorts of waste to boost sustainable production. Also included is detailed information on the system architecture for Industry 4.0. This model was tested for a year at a cement plant to see how it performed. Findings showed that total production increased by 13.24%, process waste decreased by 12.79%, overall equipment effectiveness (OEE) increased by 12.94%, total downtime decreased by 30.58%, mean time between failures (MTBF) increased by 24.68%, mean time to repair (MTTR) decreased by 25.58%, customer complaints decreased by 30%, rejection decreased by 71.7%, and, finally, specific energy consumption decreased by 71.7%.

In Jung et al. [8], the authors attempted to apply various ML algorithms for quality prediction in injection molding production. Two major elements driving Industry 4.0 are automated data gathering from machines and the application of ML algorithms to the obtained data for automated quality prediction or problem detection. Regression, Tree-Based, SVM, and autoencoders, among other commonly used ML techniques, were compared. The data for this study came from a huge injection machine dataset acquired from actual injection molding manufacturing at Hanguk Mold, a South Korean company. When evaluating the accuracy, precision, recall, and F1-score, the autoencoder models surpass the competition. The molding temperature, hopper temperature, injection time, and cycle time parameters were all found to have a significant impact in feature testing.

In Kumar et al. [9], the goal of the study is to create a big data analytics framework that optimizes the maintenance schedule using condition-based maintenance (CBM) and increases forecast accuracy to measure the remaining life prediction uncertainty. Based on feature engineering and a fuzzy unordered rule-based induction technique, they present a two-phase prediction-based maintenance big data analytics framework for the optimization of the maintenance schedule. They use feature engineering on the available gas turbine dataset in the first step. This introduces new variables, then focuses on observations that are much higher than the rest of the samples, before training a fuzzy classifier with the best prediction accuracy of the backward feature elimination approach. The outlier value is removed in the second step, and the target value is replaced by the predicted value of the trained fuzzy classifier. The experimental results are based on a large dataset obtained by a sophisticated gas turbine propulsion plant simulator.

In Quan et al. [10], the study introduces PCDEE-Circle, a systematic development framework that focuses on the contribution of human–robot collaborative disassembly (HRCD) to economic, environmental, and social sustainability. Perception, cognition, decision, execution, and evolution are the five phases of the PCDEE-Circle, which are reflected in one outward circle and two internal circles. A detailed enabling system for HRCD is also offered, along with a set of advanced technologies, such as the cyber-physical production system (CPPS) and artificial intelligence (AI). The systematic approaches for HRCD also consider deep reinforcement learning, incremental learning, and transfer learning. Using a case study, the paper demonstrated multi-modal perception for ABB industrial robots and the human body, as well as sequence planning for an HRCD task, and eventually realized a distance-based security approach and motion-driven control mode, using a case study. It demonstrates the feasibility and effectiveness of the proposed HRCD techniques and proves the systematic framework's functionality.

In the Industry 4.0 vision, Leng et al. [11] explores the landscape of blockchain-enabled sustainable manufacturing. From two viewpoints, namely the production system and product lifecycle management, this article examines how blockchain might overcome possible challenges to attaining sustainability. The survey begins with a review of the literature on these two views, followed by a discussion of the current state of research in blockchain-enabled sustainable manufacturing, which provides fresh light on pressing concerns related to the United Nations' Sustainable Development Goals. The authors discovered that the blockchain-enabled transformation of a sustainable manufacturing paradigm is still in the hype phase and is on its way to full acceptance. The poll concludes with a discussion of the obstacles that blockchain-enabled industrial applications face in terms of methodologies, societal barriers, standards, and legislation. The study ends with a consideration of the obstacles and societal constraints that blockchain technology must overcome in order to establish its long-term viability in the industrial and corporate worlds.

Research outlining the sustainable manufacturing strategy in terms of principles, implementation procedures, and assessment methodologies is offered by Kishawy et al. in [12]. Here the author addressed the sustainable manufacturing strategy in terms of principles, implementation strategies, and assessment methods. The needed sustainable aim is provided by the interplay of the three sustainable levels (process, product, and system). The author stated that decreasing energy consumption, restricting waste, boosting product durability, reducing environmental and health issues, improving product quality, and producing renewable energy supplies are the main objectives in establishing a sustainable manufacturing system. To attain these objectives, several prerequisites (e.g., approach, methods, data, study, and integration) are required. In addition, implementing the sustainable manufacturing strategy necessitates the use of various design elements. Design for environmental effect, design for resource use and economy, design for manufacturability, design for functionality, and design for social impact are among these considerations. Furthermore, the author stated that there are five primary steps that must be completed in order to build a successful, long-term system. Designing work practices and maintenance, process optimization, raw material substitution, implementing new technologies, and developing new product designs are among these steps.

By mapping and summarizing existing research efforts, identifying research agendas, and identifying gaps and opportunities for research development, the systematic review in Machado et al. [13] aims to identify how sustainable manufacturing research contributes to the development of the Industry 4.0 agenda and for a broader understanding about the links between Industry 4.0 and sustainable manufacturing. This paper contributes to Industry 4.0 research by demonstrating how sustainable manufacturing concepts and the use of new technologies can enable Industry 4.0 to have positive impacts on all sustainability dimensions in an integrated manner, as well as supporting the implementation of the Industry 4.0 agenda in the following areas: developing sustainable business models; sustainable and circular production systems; and sustainable supply chains. The findings indicate that the area is genuine but not consolidated, and that it is changing as a result of the emergence of new business models and the integration of value-creation chains. The findings also allowed for the creation of a research agenda and scenario for the field's future growth, with a focus on more normative studies on the procedures of implementing the Industry 4.0 agenda. This study is not intended to give an in-depth analysis on specific subjects, and its limitations stem from the small number of publications examined, the decision not to apply statistical analysis, and the failure to explore cross-disciplinary problems that may be generated by other academic disciplines.

2.3 SUSTAINABLE MANUFACTURING

Sustainability has been defined in a variety of ways and can mean different things to different people. Sustainable development is a significant goal in human development since sustainability is becoming an increasingly crucial prerequisite for human activities [14]. At its core, sustainable development is the belief that in the development process, social, economic, and environmental concerns should all be addressed simultaneously and comprehensively. Engineering, manufacturing, and design are just a few of the industries where the notion of sustainability has been used. The topic of sustainability is becoming increasingly important to manufacturers. Sustainable manufacturing arose from the notion of sustainable development, which was coined in the 1980s to address concerns about the environment, economic development, globalization, inequity, and other aspects [15]. Recognizing the relationship between manufacturing operations and the natural environment, for example, has become a critical aspect in industrial societies' decision-making [16]. Production process comparisons for the volume/variety matrix of the goods have traditionally been included in manufacturing plans. Today, manufacturing strategies often take into account goods and processes, as well as other characteristics such as habits, in order to combine organizational and philosophical elements into the strategy. Environmental sustainability, economic sustainability, and social sustainability are the three pillars of this approach. As part of a manufacturing supply chain, environmental sustainability entails reducing carbon footprints, water usage, non-compostable packaging, and wasteful operations. If beneficial adjustments are made to this pillar, it can save money for many firms while also contributing to environmental sustainability. Economic sustainability, in turn, entails giving businesses and other organizations incentives while still adhering to all sustainability rules. To be successful throughout time and be prepared for the future, a company must be profitable and generate enough revenue. Companies should adapt and make a profit by pursuing a sustainable approach, rather than producing money in any way and at any cost. The protection of people's health from pollution, fair working conditions, and fundamental access to resources without compromising the quality of life and fair remuneration are all examples of social sustainability. This pillar also includes education, such as teaching people about sustainability and the effects it has on them individually, as well as the hazards it might bring. It all boils down to training the next generation for success while also keeping a healthy lifestyle. Without realizing it, each of us, including you… has a significant impact on the existing and future environmental and social conditions. We must adapt, embrace change, and recognize that sustainability is the path to a prosperous future.

2.3.1 Need for sustainable manufacturing

Sustainability is good for everyone's future. It has numerous short- and long-term benefits, and our Earth's ecology cannot function without more sustainable measures being taken. To exist, all living beings rely on the resources of the planet Earth. Consumers, brands, and corporations will ruin and deplete the resources available in a matter of just a few decades if they continue to pollute and overuse them. Sustainable business methods are gaining traction among manufacturers, who are seeing considerable financial and environmental benefits. Sustainable manufacturing aims to help companies increase revenue and worldwide leadership while preserving natural resources and protecting the environment. Technology investment is critical to sustainable manufacturing—automation [17, 21], robotics, software, information technology, and sensor technologies—all of which are the foundation for the IoT [18] and required for sustainable manufacturing processes. Every step of the manufacturing process, from design to production to shipment, must be examined to secure sustainability on a regular basis. This includes packing, which is often regarded as a last-minute consideration. Improved operational efficiency through cost and waste reduction, long-term business viability and success, lower regulatory compliance costs, improved sales, and brand recognition, greater access to financing and capital, and easier employee hiring, and retention can all be regarded as key benefits of sustainability. For a variety of reasons, the implementation of sustainable manufacturing is more crucial than ever. To begin with, sustainable manufacturing is not only good for the environment [19]; it may also improve the safety of your facility, your employees, your products, and your community as a whole. By making items that are more environmentally friendly, you can ensure that the people who make and consume them are also safer. Manufacturing that is environmentally friendly can also be very cost-effective. You may lower energy usage, reduce manufacturing time, reduce waste, and use less materials by enhancing the efficiency of your equipment and processes, all of which can save money for both you and your customers. More efficient, automated equipment can also improve worker safety by shielding workers from the perils of traditional, prone-to-failure technology. Furthermore, adding sustainable manufacturing techniques into operations can help to improve your company's brand, increase customer trust, and even generate new leads. With the rising focus on sustainability around the world, many consumers are looking for businesses that use environmentally friendly manufacturing processes [20]. Customers want to feel good about the things they buy and to know that the businesses they support reflect their values and standards. Sustainable manufacturing enables clients to interact with your business and establishes trust that you share their values. Furthermore, implementing eco-friendly methods indicates your commitment to the community in which your workers live and work, demonstrating that you're serious about safeguarding the environment. There is a climate catastrophe, and a current movement is underway to transition toward a more sustainable way of living, operating, and working. Sustainability must be prioritized since our current lifestyles are insufficient. We need to be pushing the envelope, putting in additional effort, and appreciating how amazing the planet Earth is.

2.3.2 The role of IoT in sustainable manufacturing

We will be able to alter our economy by unleashing the power of IoT, making production more efficient and sustainable, and assisting economies in meeting their lofty decarbonization goals [18, 21]. In this context, the Industrial Internet of Things (IIoT) is critical because it serves as a vital enabler for sustainable production, ensuring economic efficiency while also assisting in the achievement of broad environmental goals. The key advantage of combining IoT and IIoT for sustainable production is the ability to collect and operate with precise data. Connect, monitor, and analyze your whole production process with a well-thought-out, vertically integrated IoT solution that includes software, automation, and sensors. Industries may begin to monitor, compare, and evaluate existing input in order to increase output and make it both more efficient – and environmentally friendly [21]. Digital twins of production and goods may aid in the simulation of "what-if?" situations and the selection of the best method that balances economic and environmental sustainability. In 2018, the industrial sector accounted for 37% of worldwide energy consumption. When we consider the immense issues we face, it is evident that industrial manufacturing energy efficiency will play a critical part in achieving ambitious sustainability targets. Manufacturers may use IoT-based solutions to optimize their entire energy system, making it more efficient and sustainable. One of the major advantages of IoT manufacturing, as previously said, is the ability to gather and process data in real time [22]. Users may readily spot abnormalities and patterns that cause your manufacturing process to squander or lose energy by connecting, monitoring, and comparing this set of data – and take appropriate action right immediately. Intelligent gadgets and sensors on machines, for example, can determine when they are needed independently, preventing over- or underuse and, as a result, saving energy and lowering your environmental imprint. A well-thought-out IoT manufacturing solution can let businesses connect, share, and analyze all of the data generated by your manufacturing process, allowing you to make better use of your available resources. One can start using what you have more consciously if they know exactly how many resources the company has – such as raw materials, commodities, and construction materials. Reduced usage of fluids, particularly oil and water, is one example of sustainable manufacturing that can be expanded to include many more. Allow the IoT solution to assist businesses in determining which materials or spare parts to employ, reducing over- or underproduction, and repurposing resources or waste. This not only implies less consumption, but also means longer lifespans for industrial equipment and tools, resulting in a leaner and more productive production line. In a circular economy, resource sustainability extends across the whole product lifespan, from product design to operations to breakdown and recycling of the product's parts. When the IoT is used in manufacturing and other industrial contexts, it is sometimes referred to as the Industrial IoT (IIoT for short). This helps to distinguish between the IoT applications that are particular to manufacturing and industry in general [23], and the many other IoT application groups, which include healthcare, smart home appliances, automotive, and more. The IIoT designation also distinguishes between the types of data created by and shared between devices in industrial contexts, as opposed to data related to healthcare, the status of household appliances, and so on [24]. Surprisingly, the underlying technologies of the IoT and IIoT are frequently the same, especially in terms of sensors, wireless connectivity, network routing, and other aspects.

2.3.3   Role of AI in sustainable manufacturing

Industry 4.0 is built on data in all of its forms, and it has become a vital component in many facets of advanced manufacturing [29]. The term "Industry 4.0" refers to a wide range of technological, organizational, and societal developments that are occurring across the whole value chain of industrial businesses [27]. Industry 4.0 promises to decrease development cycles and increase flexibility and customization capabilities while increasing efficiencies [28]. The prospects for leveraging data in conjunction with ubiquitously available data storage and processing capabilities and novel advanced techniques [30], such as ML, are vast as the deployment of linked machines in industrial environments grows. Data is generated at various levels and granularities during the manufacturing process, including IoT sensors such as vibration, temperature, and camera-based sensors, as well as higher-level process-related data such as work orders and part movements. Approaches for collecting knowledge and using this data to guide business decisions are becoming increasingly important. Data science, ML, and DL are among examples of AI approaches that can be used to improve goods, processes, and services [25]. These technologies assist us in gaining a deeper understanding of situations and making more precise predictions. Without explicit programming, ML explains the extraction of knowledge and insights based on experience and available data [26]. Patterns and regularities should be recognized and used to make predictions automatically. Deep learning is a collection of ML algorithms that use neural networks with numerous hidden layers to perform tasks such as picture classification, audio recognition, and language understanding, to name a few. Because AI infrastructure and applications are significantly more sophisticated than traditional IT systems, designing the correct setup for the AI system is critical. The integration of numerous components, such as data sources, batch and streaming processing, data management systems, model management, and device management, are all requirements. Edge computing is becoming increasingly important in order to meet these demands. Research on AI in the industry can be categorized into four major areas: (1) predictive quality and maintenance; (2) generative design; (3) supply chain activities; (4) human–robot collaboration.

2.4 INDUSTRY 4.0

In a connected world of big data, people, processes, services, systems, and IoT-enabled industrial assets, Industry 4.0 refers to the gathering, utilization, and analysis of actionable data and information as a technique of achieving smart industries and ecosystems of industrial innovation and collaboration [31]. Industry 4.0 is changing the way businesses make, enhance, and distribute their goods. Manufacturers are incorporating new technology into their manufacturing facilities and processes, such as the IoT, cloud computing and analytics, and AI and ML.

Advanced sensors, embedded software, and robots are used in these smart factories to gather and analyze data, allowing for better decision-making. When data from manufacturing operations are coupled with operational data from ERP, supply chain, customer service, and other business systems, new levels of visibility and insight are produced from previously isolated data [32]. Increased automation, predictive maintenance, self-optimization of process improvements, and, most importantly, a new level of efficiency and responsiveness to consumers not before feasible are all benefits of digital technology.

Smart factories that use high-tech IoT devices have increased production and better quality [33]. Using AI-powered visual insights to replace manual inspection business models decreases production mistakes and saves money and time. Quality control staff may set up a smartphone connected to the cloud with minimum expense to monitor production operations from almost anywhere. Manufacturers can spot mistakes sooner rather than later, when repair work is more expensive, by using ML algorithms.

The principles and technologies of Industry 4.0 may be used in a variety of industries, including discrete and process manufacturing, as well as oil and gas, mining, and other sectors.

2.4.1 IoT in Industry 4.0

Smart factories rely heavily on the IoT. On the manufacturing floor, sensors with an IP address are installed, allowing the machines to communicate with other web-enabled equipment [34]. Large volumes of important data can be collected, analyzed, and distributed thanks to this mechanization and connectedness.

Advanced technologies such as AI, ML, and the IoT are altering the industrial business at a rapid pace. This digital transformation has greatly improved product quality and decreased process and equipment downtime. In this procedure, the IIoT and automation are critical [35].

With its unique properties, the IIoT is causing huge disruptions in industrial applications. It greatly improves factory operating efficiency and workflows by monitoring assets and processes in real time. The IIoT opens a slew of possibilities for industrialists to improve their businesses tenfold.

Interconnectivity, automation, and real-time data monitoring and exchange are all aspects of Industry 4.0, which aspires to make the industry smarter. Because Industry 4.0 is fully client-centric, producers will have to put in extra effort to provide valuable consumer experiences and services [36]. Product visioning, product sales, production, assembly, and service management are all covered by Industry 4.0's customer-centric approach. IoT makes this procedure a breeze by assisting the owner in staying current with the always-changing demands and expectations of customers.

IoT is responsible for Industry 4.0's lightning-fast expansion, in which everything is connected to a shared network, and processes are mainly automated, obviating the need for much human participation. Real-time data monitoring aids decision-making, and predictability aids in reducing the likelihood of future dangers in the sector, as well as asset management for future workability. The following are the stages of IoT and Industry 4.0 evolution:

1. Getting things connected
2. Generating insights
3. Improving the efficiency of operations and procedures
4. Innovation

The IoT's main purpose is to make everything smart, be it a house, a building, healthcare, or a factory. Industry 4.0 encompasses cybersecurity, augmented reality, autonomous robots, digital twins, cloud computing, Big Data, linked devices, and heavy machinery. Connected devices, smart factory grids, and heavy machinery are the primary junction points of Industry 4.0 and IoT. IoT is extending the concept of manufacturing excellence by improving and playing a crucial role in every industrial function. Industry-specific IoT is the best-fit technology because it integrates information and operational technology for processes to offer real outcomes.

2.4.2 AI in Industry 4.0

Manufacturing companies utilize AI and ML to fully use the vast amounts of data generated not just on the manufacturing floor, but also throughout their many business divisions, as well as from partners and third-party sources [37]. Artificial intelligence and machine learning may be used to give insights into operations and business processes, allowing for visibility, predictability, and automation [38]. Industrial machinery, for example, is prone to failure throughout the manufacturing process. Businesses may use the data collected by these assets to do predictive maintenance with machine learning algorithms, increasing uptime, and efficiency.

AI has resulted in a shift in the way businesses operate, owing to a new type of human–machine interaction. Intelligent factories, which have given rise to Industry 4.0, are characterized by cloud-based interactions between humans and cyber-physical systems.

Intelligent factories absorb automated structures and feature digital enablers that enable machines to connect with the factory's overall systems via an IoT setup. These abilities are in high demand by industries across the board, as they attempt to maintain the competitiveness of their manufacturing units in an increasing technology environment [39].

Business intelligence is sharpened by AI, which is a significant advancement for the global economy. ML and deep learning (DL) are AI approaches that, when properly implemented, have a considerable beneficial impact on a company's return on investment (ROI). By incorporating predictive maintenance systems into manufacturing processes and replacing visual inspections with robots or cobots that perform quality controls exponentially more correctly and effectively, automatic learning substantially enhances product quality [39].

Furthermore, ML develops advanced algorithms that enable 'Smart Manufacturing,' in which data collected during production is analyzed and modifications are made automatically. DL, a subclass of ML, builds its neural networks to allow for unsupervised learning, extending the methods' autonomy even further [40]. These AI approaches result in four major advantages for Industry 4.0:

1. Production optimization
2. Supply chain integration
3. Company's adaptation to the market
4. Better product development

The intricacy of AI application in Industry 4.0, on the other hand, necessitates collaboration with professionals to develop relevant and personalized solutions. The expense of developing the requisite technology is quite costly, and it necessitates extensive internal and technical expertise.

2.5 THE BENEFITS OF USING IOT IN INDUSTRIES

2.5.1 Increased efficiency

One of the major advantages of IoT is increased efficiency. It is capable of increasing operational efficiency through the optimization of industrial processes. It can also automate, increasing efficiency and streamlining factory operations. Sensors implanted in production assets are used to track their performance in order to adjust and enhance them as needed [41].

2.5.2 Predictive maintenance

The performance and usability of assets have a significant impact on industrial production. Predictive maintenance facilitated by IoT adoption can assist process managers in forecasting and responding to an asset's workability to avoid long-term harm to production and operations. IoT sensors installed in manufacturing assets track their performance in real time and warn the management if a problem is discovered [42]. These flaws are fixed as soon as possible, saving the organization a significant amount of money.

2.5.3 Real-time data monitoring

The real-time functioning and performance of the assets may be monitored, allowing for necessary modifications in the process to boost product output and quality. Furthermore, real-time data monitoring aids in decision-making and increases factory operating efficiency [39].

2.5.4 Reduces cost

Predictive maintenance and real-time data monitoring characteristics of IoT make considerable contributions to cost reduction by enabling machines to complete activities without human supervision. As the amount of human interaction is reduced, mistakes are also reduced, lowering the cost [39].

The 4th Industrial Revolution has drastically altered our perceptions of things in the workplace. At a rapid rate, capitalists are becoming more interested in sophisticated ideas [42]. The IoT is making a big contribution to the process of making industries smarter and improving their operations.

2.6 BENEFITS OF USING AI IN INDUSTRIES

The IoT is already affecting our personal and professional lives as the world becomes increasingly automated. With all the benefits given by Industry 4.0, nowhere is this more evident than in the application of AI and robots in manufacturing industry [43]. When it comes to topics like miniaturization and precision measurements, AI's capabilities are well above human power, and it provides significantly greater quality assurance. When used correctly, AI provides several benefits for the industrial business [44].

2.6.1 Direct automation

All IoT-enabled devices are connected to the factory floor via IIoT, which integrates industrial processes with Big Data and makes them programmable via a logic controller. Because of the increased usage of precise sensing equipment, data can now be created, collected, and analyzed for all parts of the manufacturing process, from temperature to item selection and packing. AI-capable programmable logic controllers with DL capabilities may then respond automatically to the smoothly created data and make changes to the tiniest function without the need for human interaction. AI-processed

Big Data analytics may significantly increase performance throughout the whole manufacturing process and can be controlled remotely [45].

2.6.2 Continuous (24×7) production

Humans are biological entities who require routine upkeep, such as food and sleep. To keep a production plant running around the clock, shifts must be implemented, with three human workers working every 24 hours. Robots don't get tired or hungry, and they can operate on the assembly line 24 hours a day, seven days a week. This enables the growth of manufacturing capacity, which is becoming increasingly important to fulfill the expectations of global consumers. In addition, robots are more efficient in a variety of sectors, including the assembly line and the picking and packaging departments. They can drastically cut turnaround times in a variety of aspects of the organization [45].

2.6.3 Safety

Humans are flawed and prone to make errors, particularly when fatigued or preoccupied. Errors and mishaps do happen on the factory floor and in any building or processing setting; this is a problem that AI and robotic aid can almost eliminate. Remote access control necessitates a reduction in human resources, particularly when the activity is hazardous or demands extraordinary effort. Even normal working conditions will reduce the number of industrial accidents and contribute to an overall increase in safety. Furthermore, the integration of modern sensing equipment with IIoT devices makes the construction of safety guards and barriers a simpler and more effective way to safeguard human life [45].

2.6.4 Lower operational costs

Many businesses are wary of introducing AI into the manufacturing industry since it necessitates a significant financial commitment. On the other hand, the ROI is substantial and continues to grow over time. Businesses will profit from significantly reduced operating costs if intelligent machines begin to take over the regular duties of a manufacturing floor, with predictive maintenance also helping to reduce machine downtime.

Consumer desire for unique, personalized, or customized items is rising these days, although they continue to seek the greatest value. It is becoming easier and cheaper to address these demands thanks to developments such as 3D printing and IIoT-linked devices, and integrating VR or AR design approaches means the entire production process will be more cost-effective. By combining ML with CAD, systems may be built and evaluated in a virtual environment before going into production, thereby lowering the cost of trial-and-error machine testing [45].

2.6.5 Greater efficiency

The IIoT allows for the collection of large volumes of data as well as powerful analytics that may be utilized to acquire insights into customer behavior. Trends, patterns, and market changes may all be forecasted throughout time, socioeconomic sectors, and geographic marketplaces, as well as political events, macroeconomic cycles, and even weather patterns. AI can anticipate information, optimize procedures, and track inconsistencies down the supply chain from source to completed product, thanks to ML. This is supported, in particular, by technology like RFID tracking, which allows items to be monitored without the use of a physical procedure like a barcode reader [45].

2.6.6 Quality control

AI may also be used to do preventative maintenance on machines and equipment. Machines may learn to foresee malfunctions and failures using sensors to track performance and operating conditions and take action to prevent them before they happen. This can lead to speedier feedback, which can help businesses avoid unplanned downtime.

Sensors can also identify the tiniest faults, scanning them at resolutions far beyond human eyesight, resulting in increased productivity and a higher percentage of things passing quality control. Many mundane tasks are also sped up by AI, and accuracy is greatly improved. This eliminates the need for time-consuming and sometimes inaccurate quality control and in-process inspection by humans [45].

2.6.7 Quick decision making

Companies may share simulations, confer on industrial activities, and transmit crucial or significant information in real time when IIoT is combined with cloud computing and virtual or augmented reality, regardless of geographical location. Sensor and beacon data is used to identify customer behavior, allowing businesses to predict future demands and make quick manufacturing choices, as well as speeding up communication between manufacturers and suppliers [45].

2.7 CONCLUSION

The purpose of this chapter is to give an in-depth study into the fields of IoT and autonomous systems to allow the architecture of ecological production for "Industry 4.0," which refers to the advancement of production techniques and optimization. The IoT, cloud technology, AI, advanced analytics, and analysis of data are all part of the fourth phase of Industrialization. There were several considerable developments in the aforesaid sectors, and some researchers have anticipated how AI technology and the Internet of Things, when merged, would not only stimulate sustainability but also benefit the production sector. Integrating AI subdisciplines like ML, computational linguistics (NLP), and computer vision might aid enterprises in developing better energy-efficient operations. The whole industrial cycle may be optimized when AI is integrated with edge computing approaches, in which information is recorded, analyzed, and maintained instantly at IoT terminals. Furthermore, the implementation of "Industry 4.0" allows for an even better fuel-efficient and ecologically beneficial strategy, as well as a reduction in massive firms' carbon footprint. The outcome of developing and ensuring effective industrial production can be described as energy-efficient, versatile to the rapidly increasing requirements of the customer and prospective clients and providing the capability to react dynamically to the manufacturing inadequacies. Amongst the various recommended methodologies, we have extensively reviewed the architectures with the most promising features.

REFERENCES

[1] Ashton, Thomas Southcliffe. "The industrial revolution 1760–1830." *OUP Catalogue*, 1997.

[2] De Vries, Jan. "The industrial revolution and the industrious revolution." *The Journal of Economic History* 54, no. 2 (1994): 249–270.

[3] Lasi, Heiner, Peter Fettke, Hans-Georg Kemper, Thomas Feld, and Michael Hoffmann. "Industry 4.0." *Business & Information Systems Engineering* 6, no. 4 (2014): 239–242.

[4] Geum, Youngjung, Hongseok Jeon, and Hakyeon Lee. "Developing new smart services using integrated morphological analysis: Integration of the market-pull and technology-push approach." *Service Business* 10, no. 3 (2016): 531–555.

[5] Ghobakhloo, Morteza. "Industry 4.0, digitization, and opportunities for sustainability." *Journal of Cleaner Production* 252 (2020): 119869.

[6] Enyoghasi, Christian, and Fazleena Badurdeen. "Industry 4.0 for sustainable manufacturing: Opportunities at the product, process, and system levels." *Resources, Conservation and Recycling* 166 (2021): 105362. ISSN 0921-3449. https://doi.org/10.1016/j.resconrec.2020.105362

[7] Jena, M.C., S.K. Mishra, and H.S. Moharana. "Application of Industry 4.0 to enhance sustainable manufacturing." *Environmental Progress & Sustainable Energy* 39 (2020): e13360. https://doi.org/10.1002/ep.13360

[8] Jung, H., J. Jeon, D. Choi, and J.-Y. Park. "Application of machine learning techniques in injection molding quality prediction: Implications on sustainable manufacturing industry." *Sustainability* 13, no. 8 (2021): 4120. https://doi.org/10.3390/su13084120

[9] Kumar, Ajay, Ravi Shankar, and Lakshman Thakur. "A big data-driven sustainable manufacturing framework for condition-based maintenance prediction." *Journal of Computational Science* 27 (2017). https://doi.org/10.1016/j.jocs.2017.06.006

[10] Quan, Liu, Liu Zhihao, Xu Wenjun, Tang Quan, Zhou Zude, and Pham Duc Truong. "Human-robot collaboration in disassembly for sustainable manufacturing." *International Journal of Production Research* 57, no. 12 (2019): 4027–4044. https://doi.org/10.1080/00207543.2019.1578906

[11] Leng, Jiewu, Guolei Ruan, Pingyu Jiang, Kailin Xu, Qiang Liu, Xueliang Zhou, and Chao Liu. "Blockchain-empowered sustainable manufacturing and product lifecycle management in industry 4.0: A survey." *Renewable and Sustainable Energy Reviews* 132 (2020): 110112.

[12] Kishawy, Hossam A., Hussien Hegab, and Elsadig Saad. "Design for sustainable manufacturing: Approach, implementation, and assessment." *Sustainability* 10, no. 10 (2018): 3604.

[13] Machado, Carla Gonçalves, Mats Peter Winroth, and Elias Hans Dener Ribeiro da Silva. "Sustainable manufacturing in Industry 4.0: An emerging research agenda." *International Journal of Production Research* 58, no. 5 (2020): 1462–1484.

[14] Christian, Enyoghasi, and Fazleena Badurdeen. "Industry 4.0 for sustainable manufacturing: Opportunities at the product, process, and system levels." *Resources, Conservation and Recycling* 166 (2021): 105362. ISSN 0921-3449. https://doi.org/10.1016/j.resconrec.2020.105362

[15] Ming-Lang, Tseng, Thi Phuong Thuy Tran, Hien Minh Ha, Tat-Dat Bui, and Ming K. Lim. "Sustainable industrial and operation engineering trends and challenges Toward Industry 4.0: A data driven analysis." *Journal of Industrial and Production Engineering* 38, no. 8 (2021): 581–598. https://doi.org/10.1080/21681015.2021.1950227

[16] Surajit, Bag, Jan Ham Christiaan Pretorius, Shivam Gupta, and Yogesh K. Dwivedi. "Role of institutional pressures and resources in the adoption of big data analytics powered artificial intelligence, sustainable manufacturing practices and circular economy capabilities." *Technological Forecasting and Social Change* 163 (2021): 120420. ISSN 0040-1625. https://doi.org/10.1016/j.techfore.2020.120420

[17] Jamwal, A., R. Agrawal, M. Sharma, A. Kumar, V. Kumar, and J.A.A. Garza-Reyes. "Machine learning applications for sustainable manufacturing: A bibliometric-based review for future research." *Journal of Enterprise Information Management* (2021). https://doi.org/10.1108/JEIM-09-2020-0361

[18] Mastos, T.D., A. Nizamis, T. Vafeiadis, N. Alexopoulos, C. Ntinas, D. Gkortzis, A. Papadopoulos, D. Ioannidis, and D. Tzovaras. "Industry 4.0 sustainable supply chains: An application of an IoT enabled scrap metal management solution." *Journal of Cleaner Production* (2020). https://doi.org/10.1016/j.jclepro.2020.122377

[19] Jamwal, Anbesh, Rajeev Agrawal, Monica Sharma, and Antonio Giallanza. "Industry 4.0 technologies for manufacturing sustainability: A systematic review and future research directions." *Applied Sciences* 11, no. 12 (2021): 5725. https://doi.org/10.3390/app11125725

[20] Routray, S.K., K.P. Sharmila, A. Javali, A.D. Ghosh, and S. Sarangi. "An Outlook of Narrowband IoT for Industry 4.0." *2020 Second International Conference on Inventive Research in Computing Applications (ICIRCA)*, 2020, pp. 923–926. https://doi.org/10.1109/ICIRCA48905.2020.9182803

[21] Du Preez, Anli and Gert Oosthuizen. "Machine learning in cutting processes as enabler for smart sustainable manufacturing." *Procedia Manufacturing* 33 (2019): 810–817. https://doi.org/10.1016/j.promfg.2019.04.102

[22] Papetti, Alessandra, Fabio Gregori, Monica Pandolfi, Margherita Peruzzini, and Michele Germani. "IoT to enable social sustainability in manufacturing systems." *Advanced Transdisciplinary Engineering* 7 (2018): 53–62. https://doi.org/10.3233/978-1-61499-898-3-53

[23] Dastbaz, Mohammad, and Peter Cochrane. *Industry 4.0 and Engineering for a Sustainable Future*, 2019. https://doi.org/10.1007/978-3-030-12953-8

[24] Li, K., T. Zhou, and Bh. Liu. "Internet-based intelligent and sustainable manufacturing: Developments and challenges." *International Journal of Advanced Manufacturing Technology* 108 (2020): 1767–1791. https://doi.org/10.1007/s00170-020-05445-0

[25] J. Lee, H. Davari, J. Singh, and V. Pandhare. "Industrial artificial intelligence for Industry 4.0-based manufacturing systems." *Manufacturing Letters* (2018). https://doi.org/10.1016/j.mfglet.2018.09.002

[26] Çınar, Zeki M., Abubakar Abdussalam Nuhu, Qasim Zeeshan, Orhan Korhan, Mohammed Asmael, and Babak Safaei. "Machine learning in predictive maintenance towards sustainable smart manufacturing in Industry 4.0." *Sustainability* 12, no. 19 (2020): 8211. https://doi.org/10.3390/su12198211

[27] Kishawy, H.A., H. Hegab, and E. Saad. "Design for sustainable manufacturing: Approach, implementation, and assessment." *Sustainability* 10 (2018): 3604. https://doi.org/10.3390/su10103604

[28] Rosen, M.A., and H.A. Kishawy. "Sustainable manufacturing and design: Concepts, Practices and Needs." *Sustainability* 4 (2012): 154–174. https://doi.org/10.3390/su4020154

[29] Felsberger, Andreas, Fahham Hasan Qaiser, Alok Choudhary, and Gerald Reiner. "The impact of Industry 4.0 on the reconciliation of dynamic capabilities: Evidence from the European manufacturing industries." *Production Planning and Control* 33, no. 2–3 (2022): 277–300.

[30] Sharma, Lakhan, Ravinder Kumar, Pratyaksh Tyagi, Lincon Nagar, and Devdutt Gaur. "Challenges of Sustainable Manufacturing for Indian Organization: A Study." In *Recent Advances in Mechanical Infrastructure: Proceedings of ICRAM 2019*, 2020. https://doi.org/10.1007/978-981-32-9971-9_4

[31] Lasi, Heiner, Peter Fettke, Hans-Georg Kemper, Thomas Feld, and Michael Hoffmann. "Industry 4.0." *Business & Information Systems Engineering* 6, no. 4 (2014): 239–242.

[32] Ghobakhloo, Morteza. "Industry 4.0, digitization, and opportunities for sustainability." *Journal of Cleaner Production* 252 (2020): 119869.

[33] Okano, Marcelo T. "IOT and Industry 4.0: The Industrial New Revolution." In *International Conference on Management and Information Systems*, vol. 25, p. 26, 2017.

[34] Manavalan, Ethirajan, and Kandasamy Jayakrishna. "A review of Internet of Things (IoT) embedded sustainable supply chain for industry 4.0 requirements." *Computers & Industrial Engineering* 127 (2019): 925–953.

[35] Dalenogare, Lucas Santos, Guilherme Brittes Benitez, Néstor Fabián Ayala, and Alejandro Germán Frank. "The expected contribution of Industry 4.0 technologies for industrial performance." *International Journal of Production Economics* 204 (2018): 383–394.

[36] Aheleroff, Shohin, Xun Xu, Yuqian Lu, Mauricio Aristizabal, Juan Pablo Velásquez, Benjamin Joa, and Yesid Valencia. "IoT-enabled smart appliances under industry 4.0: A case study." *Advanced Engineering Informatics* 43 (2020): 101043.

[37] Popkova, Elena G., and Bruno S. Sergi. "Human capital and AI in industry 4.0. Convergence and divergence in social entrepreneurship in Russia." *Journal of Intellectual Capital* 21 (2020): 565–581.

[38] Sergi, Bruno S., Elena G. Popkova, Aleksei V. Bogoviz, and Tatiana N. Litvinova, eds. *Understanding Industry 4.0: AI, the Internet of Things, and the Future of Work*. Emerald Group Publishing, 2019. https://www.emerald.com/insight/content/doi/10.1108/978-1-78973-311-220191002/full/html

[39] https://nexusintegra.io/artificial-intelligence-the-driving-force-behind-industry-4-0/

[40] Candanedo, Inés Sittón, Elena Hernández Nieves, Sara Rodríguez González, M. Martín, and Alfonso González Briones. "Machine Learning Predictive Model for Industry 4.0." In *International Conference on Knowledge Management in Organizations*, pp. 501–510. Springer, Cham, 2018.

[41] de Vass, Tharaka, Himanshu Shee, and Shah Miah. "IoT in supply chain management: Opportunities and challenges for businesses in early Industry 4.0 context." *Operations and Supply Chain Management: An International Journal* 14, no. 2 (2021): 148–161.

[42] Mohamed, Mamad. "Challenges and benefits of Industry 4.0: An overview." *International Journal of Supply and Operations Management* 5, no. 3 (2018): 256–265.

[43] Dhanabalan, T., and A. Sathish. "Transforming Indian industries through artificial intelligence and robotics in industry 4.0." *International Journal of Mechanical Engineering and Technology* 9, no. 10 (2018): 835–845.

[44] https://towardsdatascience.com/a-very-brief-introduction-to-ai-in-the-industry-4-0-14e6f4b46cd1

[45] https://www.rowse.co.uk/blog/post/7-manufacturing-ai-benefits

Chapter 3

Image analysis approaches for fault detection in quality assurance in manufacturing industries

Vijayakumar Ponnusamy and Dilliraj Ekambaram

SRM Institute of Science and Technology, Chennai, Tamil Nadu, India

CONTENTS

DOI: 10.1201/9781003257714-3

3.1 INTRODUCTION

Industry 4.0 is a term that is becoming very popular in manufacturing industries as a short term to describe the provision of modern industrial activities. This innovation led to a concentration on reducing the employee workload by replacing the conventional method and importing advanced technology for activities such as the inspection of materials, welding, assembling, production, maintenance, etc., in manufacturing industries. An increment emphasis on standard production requires manufacturing industries constantly to provide improved quality products with more complex, low cost and limited use of resources without having any effect on the external environment. To control environmental change and forestall a further rises in global temperatures, the European Union introduced the "European Green Deal" in 2019 [3]. One of the essential phases in the process of manufacturing is that of inspection. The term "Quality" has a unique global value in manufacturing. Manufacturing industries focus on providing customer satisfaction through quality process aspects such as Customer-–driven engineering, Design of Experiments (DoE), Statistical Process Control (SPC) tools, Acceptance sampling, Failure Mode and Effects Analysis (FMEA) and Six sigma [26].

Ensuring product quality is a most significant perspective in the modern competitive environment. At present, a number of different methodologies are used to survey the nature of an item or the result of an interaction. Depending upon the different modern techniques utilized to recognize a deformity on a surface/volume, quality control procedures are classified as either disastrous or non-disastrous [27]. Figure 3.1 shows the various strategies used for quality control in the manufacturing industries.

In progress in the development of manufacturing industries, Additive Manufacturing (AM) is an essential part of showing the overall product in the 3D model through connecting the different materials in the immersive [16]. AM is a rapidly developing technique showing progress in advanced manufacturing technologies. It has been an emerging technique since the 1980s. AM offers critical benefits contrasted with conventional subtractive assembling, such as a more elevated level of plan opportunity and improved assembling capacity, enabling the creation of specific, on-demand products. AM is compatible with various substances such as plastics, metals, composites, and resins. Figure 3.2 illustrates the different AM techniques to determine the defect in the metal [17].

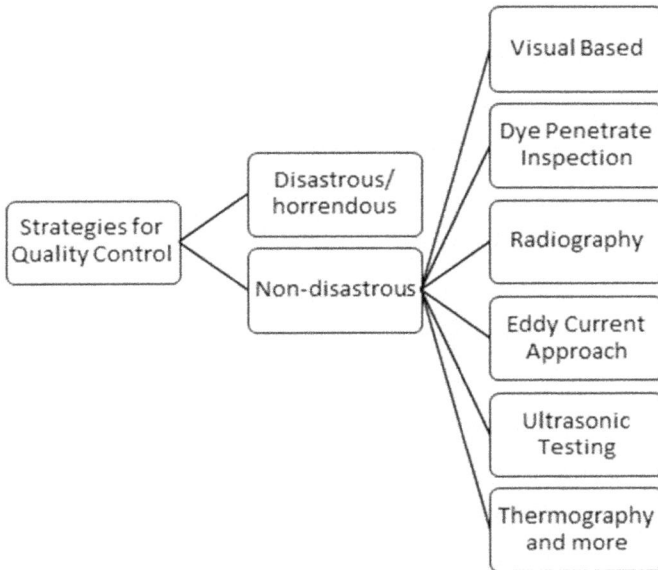

Figure 3.1 Strategies implied in quality control.

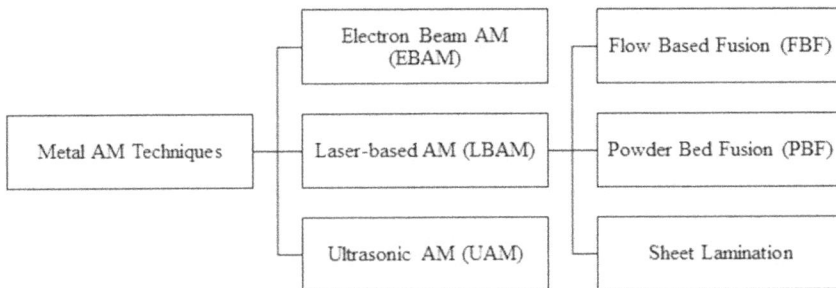

Figure 3.2 Various AM techniques for metal.

AM modeling is classified based on key performance indicators and process elements. The key performance indicators (KPIs) group the demonstrating approaches of each interaction bunch and the cycle boundaries utilized. It must be noted that the demonstrating move followed by specific examinations does not mean that the association of the cycle boundaries to the recreated KPIs. In such cases, the table cell alluding to the cycle boundaries is left vacant, featuring the KPI-focused point of view of those methodologies [25].

Figure 3.3 shows the issues in AM related to decreasing productivity, making it unable to attain the quality and improper mechanical properties of the final product.

AM Part Quality and Production Rate

People
- Work Experience
- Knowledge
  - Lack of Courses
- New Capabilities that are not fully exploited
  - Different Limitations than traditional processes
  - • New Design philosophy

Machines
- Operating Conditions
- New Technology
- Standardization
- Monitoring
- Wear
- Build Orientation
- Path Planning
- Building Speed

Processes
- Post Processing
- Thermal Distortions
  - Non-Uniform Thermal Phenomena
  - • Residual Stresses
- Phase Transitions
  - Coexistence of different states of matter
- Viscous Flow
  - Cutting
  - • Fluid Dynamics
  - • Thermal Phenomena
  - • Multiphysics

Raw Materials
- Availability of Materials
  - Delivery Times
- Moisture Content
  - Storage Conditions
- Material Properties
  - • Temperature
  - • Shelf Life
- Many Types of Materials
- Material Composition
- Quality of Materials
- Standardization of Materials

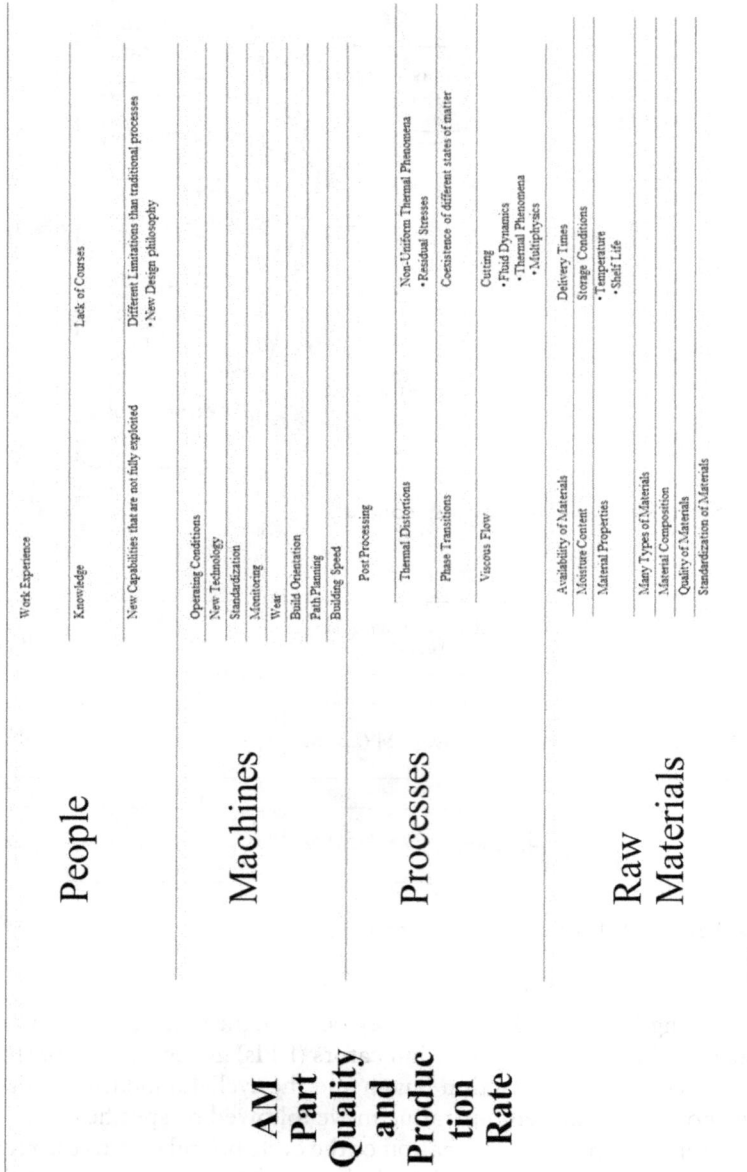

Figure 3.3 Significant issues in the AM technology.

The current economy needs new solutions for monitoring the automated production of goods and identifying abandoned materials. Careful observation of the mechanical components of innovation frameworks or machines is required for the programmed invention. Deep Learning has proven its effectiveness in feature extraction from images, videos, and text to succeed in various item finding, acknowledgment, division, and characterization projects. Despite these developments, a microscopic study has been done on the applicability of unusually designed Convolutional Neural Networks (CNNs) for deformity location and contemporary item recognition [3]. In a similar fashion, we have assessed deliberated the various modern technologies used for visual inspection in this chapter.

To focus further on refinement in the present model in an attempt to achieve more accuracy, an improvement in image acquisition timing in the existing systems helps to reduce processing time. System performance can be improved by the optimization of depth analysis in an object with smaller defects; more features, such as curvature identification and more macro-level features for individual points in 3D reconstruction on objects, need to be improved, scaling up computational power to process faster and provide improved accuracy for the real-time tracking of moving objects. Real-time augmentation is required for training and testing models so as to increase the best solutions to classify the defect and non-defect items in the industries.

The results from this study can be summed up as follows: high accuracy, high situating, fast location, small objects, complex foundation, impeded object recognition and item affiliation. In conclusion, we contour the current accomplishments and impediments of the present techniques, alongside the fluctuations of research difficulties, to help examine the local area on imperfection recognition in setting a further plan for future investigations.

3.1.1 Variety of defects and their classifications

In the production area of modern-day industries, quality control targets keeping a quality level or limiting imperfections for further maintenance. Customary identification strategies ordinarily manage standard, large-scale estimated and complex varieties of surface deformities. Every counterfeit visual deformity recognition strategy intends to identify and group defects for additional handling. Modern applications need all-around organized information bases of the conceivable deformity types in order to achieve appropriate classification. Regardless, spreading such a general and broad database for a classifier is tempting, given the anomaly and the unique nature of the disfigurements that can occur in the movement circumstances. In such a manner, a general grouping approach is rarely mentioned; every application uses a material-based distortion classifier.

The categorization of surface imperfections detection for various materials in the surfaces is categorized into two significant gatherings: visible defects and recognizable defects. Significantly, the arrangement is fundamental, basic theoretical and not suitable for methods with explicit prerequisites. Still, it gives areas of strength for a dependable reason for a grouping with an artificial consciousness framework. The crucial suspicion of this deformity arrangement is that the characterization of imperfection is a seriously abstract judgment, i.e., it incredibly relies upon what a deformity addresses for the human manager. This choice is generally founded on an edge and a realistic portrayal of the size proportion of both the part and the imperfection [27].

The surface defects classifications are displayed in Figure 3.4. The two categories of surface defects are divided further based on the significant impacts affected in the final products, such as Line Shape, Patch/Spot Shape

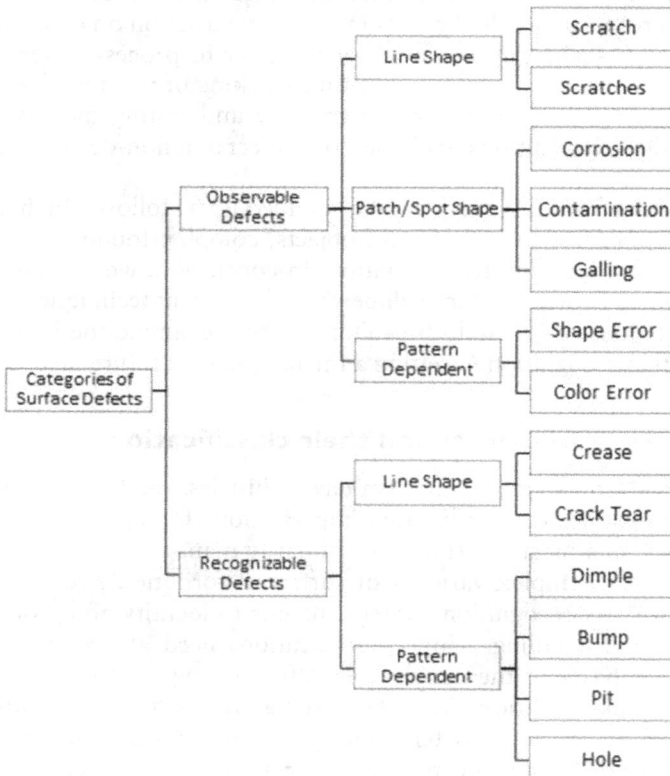

Figure 3.4 Categories of various surface defects.

and Pattern Dependent. We deliberate the further classification of these categories below.

Observable surface defects have the following common characteristics:

- It is not possible to detect the defects with tactile sensation.
- Scratch defects are more profound and broader.
- Some defects are line-shaped; others are patch-shaped.
- Defects such as corrosion and contamination change the material properties.
- Shape and color error defects occurred due to mistakes made by the workers.

Similarly, recognizable defects have characteristics such as:

- Most defects are patch-shaped: others are line-shaped defects.
- All defects are possible to detect with tactile and visual.
- Significantly deeper depth compared to object thickness.
- It occurs due to concentrated pressure on the surface.
- Some defects occur due to friction and the immediate, smooth transition of objects on the surface.

3.1.2 Summary of various machine learning algorithms

In order to make the system inspect the material proficiently, more database information is required to train the model with machine learning (ML). Hence, to concentrate on deciphering the report, a subsequent review must perform with the assistance of Artificial Intelligence (AI). Given the extensiveness of the datasets that are available, there is currently a rapidly increasing interest in AI. Enormous ventures apply AI to separate meaningful information. The reason for AI is to gain from the data. Numerous investigations have been undertaken on the most proficient method to train the model to predict deformity surface. Many mathematicians and software engineers apply a few ways to find the arrangement of this issue: gigantic informational collections.

No one single algorithm is fit for all different types of applications. Accordingly, researchers or data scientists have to choose the algorithm based on their application development. The researchers can choose which kind of classification data to focus on when training the model to predict the outcome [4]. Figure 3.5 shows the summary of various ML algorithms used, all of which are discussed in subsequent sections of this chapter.

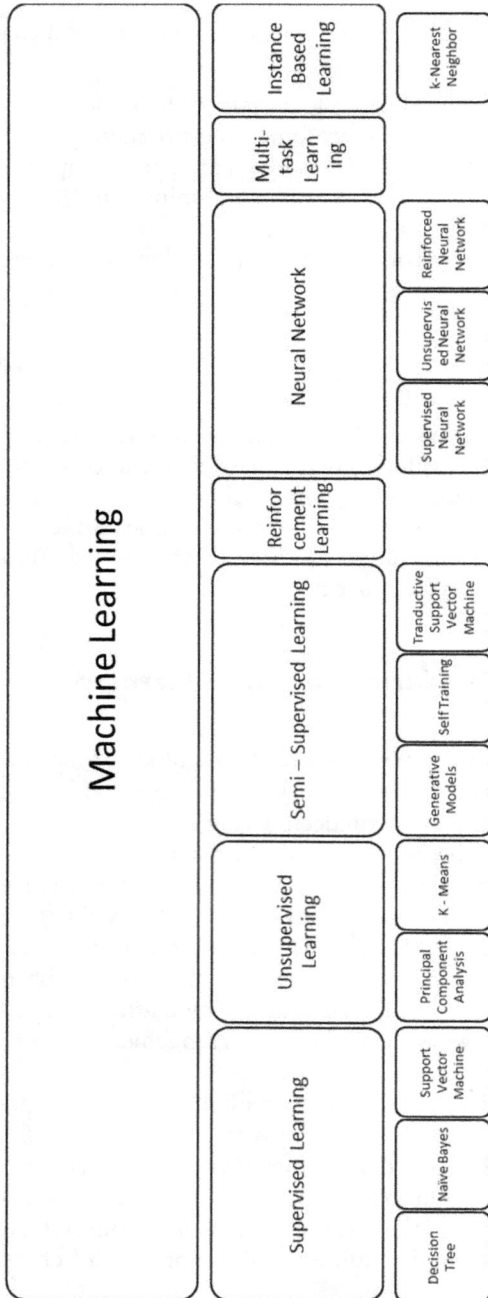

Figure 3.5 Summary of various machine learning algorithms.

Here, we deliberate the different categories of ML algorithms in detail.

Supervised Learning: This is a well-known approach for the operations such as the classification and regression of datasets. This approach is used to classify the patterns of, or similarities between, various datasets. This approach is the best for the prediction or classification of multiple datasets. Figure 3.6 shows the workflow process to detect and classify the different datasets. A few supervised ML algorithms are outlined in Table 3.1.

Unsupervised Learning: This approach derives the conclusion for predicting the raw input data using deep learning. These algorithms are left alone to decide to discover the unlabelled data. This learns the features from the input data and classifies the other data using the features of known data. Principally, it focuses on clustering performance—the algorithms in this learning are classified as principal component analysis [29] and k-means. Figure 3.7 illustrates the process of unsupervised learning.

Semi-supervised Learning: This approach combines supervised and unsupervised learning to predict input data. Figure 3.8 shows the process of semi-supervised learning.

Reinforcement Learning: This approach takes the decision, not with the help of supervised learning. It can predict results based on the reinforcement agent. The edges of the images are mostly taken as agents to classify the images. Figure 3.9 illustrates the operation of this approach.

3.1.3 Summary of various deep learning (DL) algorithms

One essential progress in the present deep learning scenario is the application of a machine vision framework to assess product quality in the industries [20] because of the wide assortment of deformity surfaces in industries, determining the small-sized surface defects with high-quality pictures in the assembling environment [18]. Some specular reflection in the material surface led our system to the wrong prediction. Hence, we must be aware of the

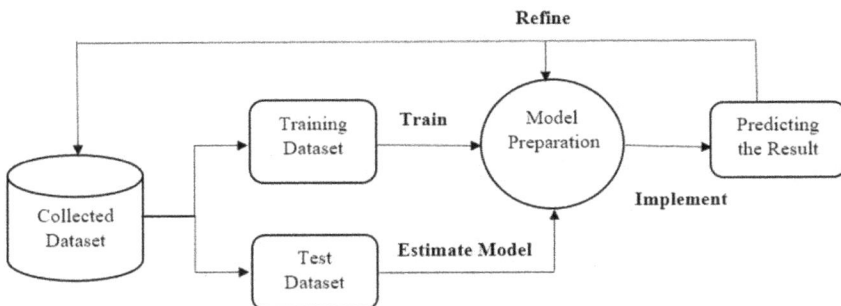

Figure 3.6 Workflow process of supervised learning.

Table 3.1 Detailed categories of supervised ML algorithms

Algorithms	Decision tree (DT)	Naïve Bayes	Support vector machine (SVM)
Outline	It is a simple real-time approach for predicting the same pattern of datasets. It can predict the data by learning the information through the training datasets.	Focuses on the message characterization industry and utilized for grouping and characterization reason relies upon the restrictive likelihood of occurring.	It is done by making the hyperplane or hyper sets in the limited or boundless layered space. It is one of the supervised learning optimizations or grouping algorithms.
Process	Learning and prediction can be made in a two-step process. The first step, learn from the training data. The second step is generating the leaf based on the available training data.	It finds the probability of sample data and predicts the object in the overall dataset. [4]	It efficiently can perform non-linear classification and indirectly match their inputs into boundless layered space. [28]

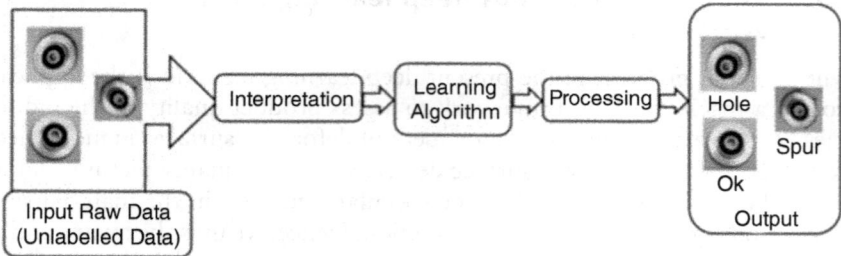

Figure 3.7 Process of unsupervised learning.

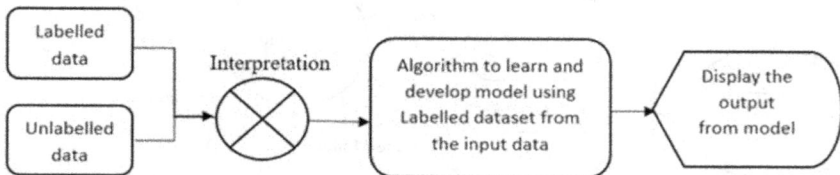

Figure 3.8 Process of semi-supervised learning.

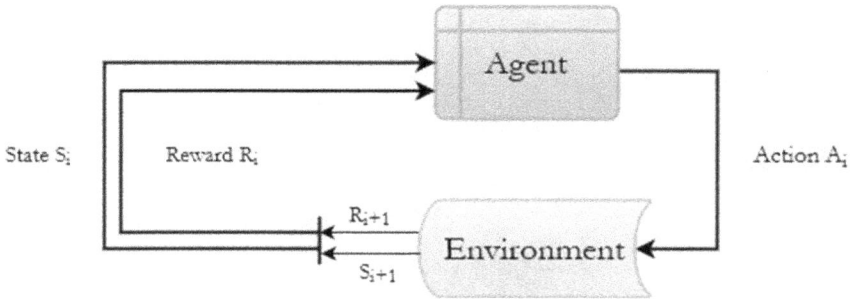

State S_i Reward R_i Action A_i

R_i+1

S_i+1

Agent

Environment

Figure 3.9 Operation of reinforcement learning.

lighting conditions before predicting the defects. Many attempts are made to identify the weaknesses by extracting the lighting on the surface, making the material shinier and unable to determine the defects properly through image capture device. These are the biggest challenges in defect detection. This section deliberates the different deep learning algorithms in brief as follows:

3.1.3.1 Convolutional neural network (CNN)

Detecting defects in the material surface requires actual progress on improving computing power and following the standard to empower the framework [21, 24]. Over the past twenty years, numerous profound brain networks have been outlined, including profound conviction organization, convolutional neural networks (CNN), and intermittent brain organization; of these, CNN is the most renowned [10]. In light of the blunder, the boundaries are changed and subsequently taken care of in the neural network. This type of neural network is utilized in straightforward neural organization. Extracting the pattern features from the input and training the model to predict the defects is another method employed in CNN. CNN is generally examined in the AM field, as it is produced for picture handling and example acknowledgment, thus appropriate for detecting pictures [17].

Three kinds of layers are regularly employed in CNN engineering: the convolutional layer, pooling layer, and fully connected layer, which are described in Figure 3.10. Of these, convolutional and pooling layers are utilized to remove spatial elements (e.g., theme shape, relative areas between themes) from the pictures. The completely associated layers connect the separated highlights to the picture classifications to be characterized.

During the creation of a neural network to improve the progression in convolution activity, we require [1]. The first and second requests limited discrete convolution activity described by the Equations (3.1) and (3.2).

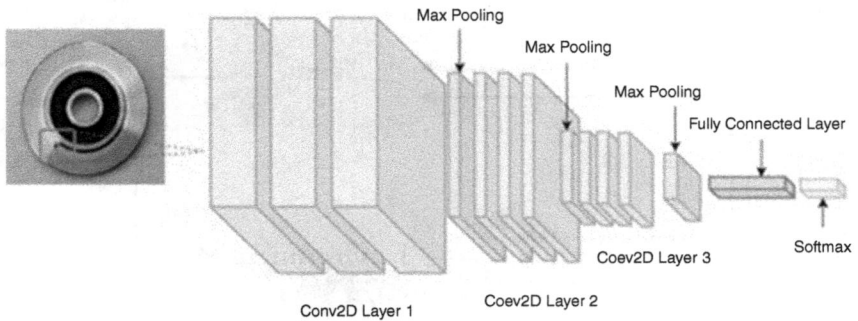

Figure 3.10 Convolutional neural network.

$$y(k) = (h * u)(k) = \sum_{n=0}^{M} h(n) u(k-n) \tag{3.1}$$

$$y(k) = (h * u)(k) = \sum_{n_1=0}^{M} \sum_{n_2=0}^{M} h(n_1, n_2) u(k-n_1) u(k-n_2) \tag{3.2}$$

where $h(n)$ and $h(n_1, n_2)$ are first-order and second-order convolution kernel, respectively. $u(k)$ is the input of the network [1].

3.1.3.2 Generative adversarial

Generative adversarial networks (GANs) comprising two profound brain organizations (i.e., a generator and discriminator) is an average semi-administered learning technique [22]. The generator constantly produces new pictures and feeds them to the discriminator. Using natural and fake classifiers, GANs are generally applied to create flawed images [23] to grow the constraint of deformity tests

3.1.3.3 Recurrent neural network (RNNs)

RNNs and feed-forward brain networks get their names from the way in which they channel data. In a feed-forward brain organization, the data moves in a single heading from the info layer, through the hidden layers, to the result layer [30]. The data moves straight through the organization and never contacts a different neurons on the same time. two times. Feed-forward brain networks retain no memory of the information they get and are very poor at anticipating what is to follow immediately. Since a feed-forward network considers ongoing communication, it has no thought of a request in time. It cannot recall anything about what occurred in the past other than its preparation.

3.1.3.4 Radial basis function networks (RBFNs)

A radial basis function network (RBFN) is a model brain network that involves outspread premise capacities as enactment capacities. The result of the organization is a straight mix of spiral premise elements of the sources of info and neuron boundaries. RBFNs have many purposes, including capacity estimate, time series expectation, grouping, and framework control. The way of behaving of the organization relies upon the loads and the enactment of an exchange work F indicated for the units [31].

3.1.3.5 Long short-term memory networks (LSTMs)

A Long Short-Term Memory Network is a high-level RNN, a consecutive organization that permits data to persevere. It is fit for taking care of the disappearing angle issue by RNN. A repetitive brain network, or RNN, is used for dynamic memory. Suppose you recall the past scene while watching a video or perusing a book; you realize what occurred in the previous section. This corresponds with the manner in which RNNs work; they reflect the past data and use it to handle the ongoing info. The drawback of RNNs is that they cannot recall long-term conditions because of the evaporating slope. LSTMs are unequivocally intended to keep away from long-haul reliance issues.

3.1.3.6 Self-organizing maps (SOMs)

This system followed an unaided learning approach and prepared its organization through a severe learning calculation. A self-organizing map (SOM) is utilized for grouping and planning (or dimensionality decrease) strategies to plan multi-layered information onto lower-layered information, allowing individuals to reduce complex issues for simple understanding. SOM has two layers; one is the Input layer, and the other one is the Output layer.

3.1.3.7 Restricted Boltzmann machines (RBMs)

The RBMs can handle the effective modeling distribution over binary valued data. Figure 3.11 indicates the exponential distribution of binary data for

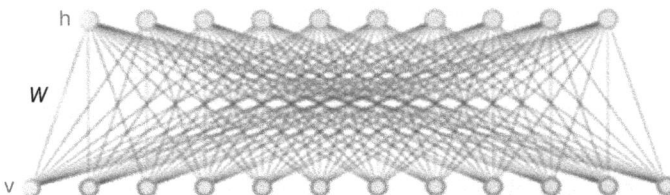

Figure 3.11 Exponential RBMs.

various applications in different domains. Here, the top layer 'h' represents the vector of "hidden" binary variables, and the bottom layer 'v' represents the vector of "visible" binary variables [31].

The RBMs can be added with Gaussian Bernoulli to support Real valued/Whole valued inputs for various applications such as speech and image classifications. The replicated SoftMax model is an extension of RBMs to count the sparse data from the image [31].

3.1.3.8 Multilayer perceptron (MLPs)

The multilayer perceptron (MLP) network is a whole-brain organization network made out of N-layered info and an M-layered yield with a hidden layer and utilizing the back spread learning model, as displayed in Figure 3.12. We used a solitary layer network with 100 hubs. The enactment capacity of the model is Rectified Linear Unit (ReLU), and Adam's streamlining agent was applied in this model.

3.1.3.9 Autoencoders

An autoencoder is a fake brain network used to learn productive coding of unlabelled information (solo learning). The encoding is approved and refined by endeavoring to recover the contribution from the encoding. The autoencoder knows a portrayal (encoding) for a bunch of information, ordinarily for dimensionality decrease, via preparing the organization to disregard irrelevant information ("commotion").

3.1.3.10 Deep belief networks (DBNs)

Deep belief networks (DBNs) are probabilistic generative models that contain many layers of stowed-away factors. Each layer catches high-request connections between the exercises of stowed away elements in the layer beneath. The main two layers of the DBN structure are an RBM model in which the lower layer's structure is a coordinated sigmoid conviction organization [31].

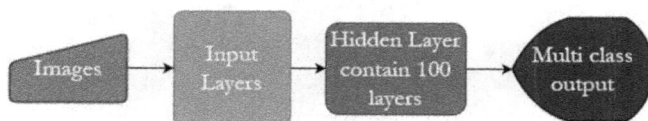

Figure 3.12 Multilayer perceptron.

3.2 VARIOUS APPROACHES TO DETECT DEFECTS IN CASTING PRODUCTS

3.2.1 Various deep learning approaches

This section discusses the various deep learning approaches and its process.

3.2.1.1 CNN with MVGG19 network

The majority of the various standard evened-out multipath plans of VGG19 are created by the Visual Geometry Group (VGG), VGG19 standard evened-out multipath plans and established a connection between early and late convolutional impedes. This methodology deals with a cerebrum association of 2500 neurons and three BD layers (group standardization, quitter, and worldwide average pooling), giving subsequent results.

The parameters used for the MVGG19 network are described below:

➢ The final convolutional layer has trainable layers.
➢ Network-wide batch normalization.
➢ 50% dropout in batch normalization blocks.
➢ Global pooling from average pooling.
➢ Pooling between convolutions from max pooling 2500 nodes neural network classifier.
➢ Adam optimizer.
➢ A typical value of batch size for training is 64.
➢ Depending on datasets, 30–60 epochs were used.

This change is intended to isolate the early and late separated features from the consecutive building and directly interface them with the classifier at the organization's top [2]. The architecture of CNN with the MVGG19 network is described in Figure 3.13.

3.2.1.2 Photometric stereo algorithm with customed segmentation network

This framework utilizes four different light sources to detect the defect in the surface material. It identifies the defected surface region by fixing the surface's coordination points and passing the light source through the material's surface. Next, using the profundity mapping, the defected area is located [18]. Figure 3.14 shows the segmentation network to detect the defect in the object's surface.

The overall framework architecture is shown in Figure 3.15. It consists of an Image Processing Module, a Database Module, a Processing Module, an Annotation Module and a Display Layer.

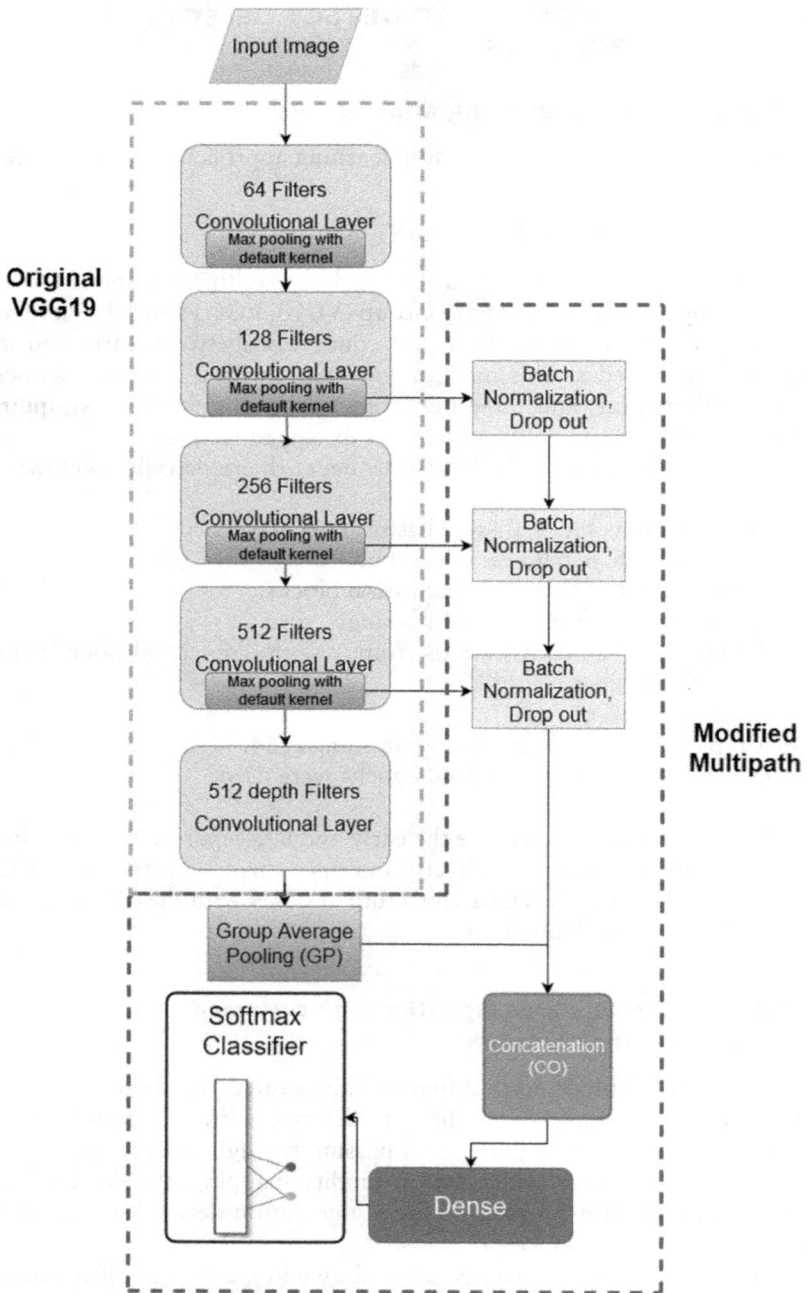

Figure 3.13 The architecture of CNN with MVGG 19 network.

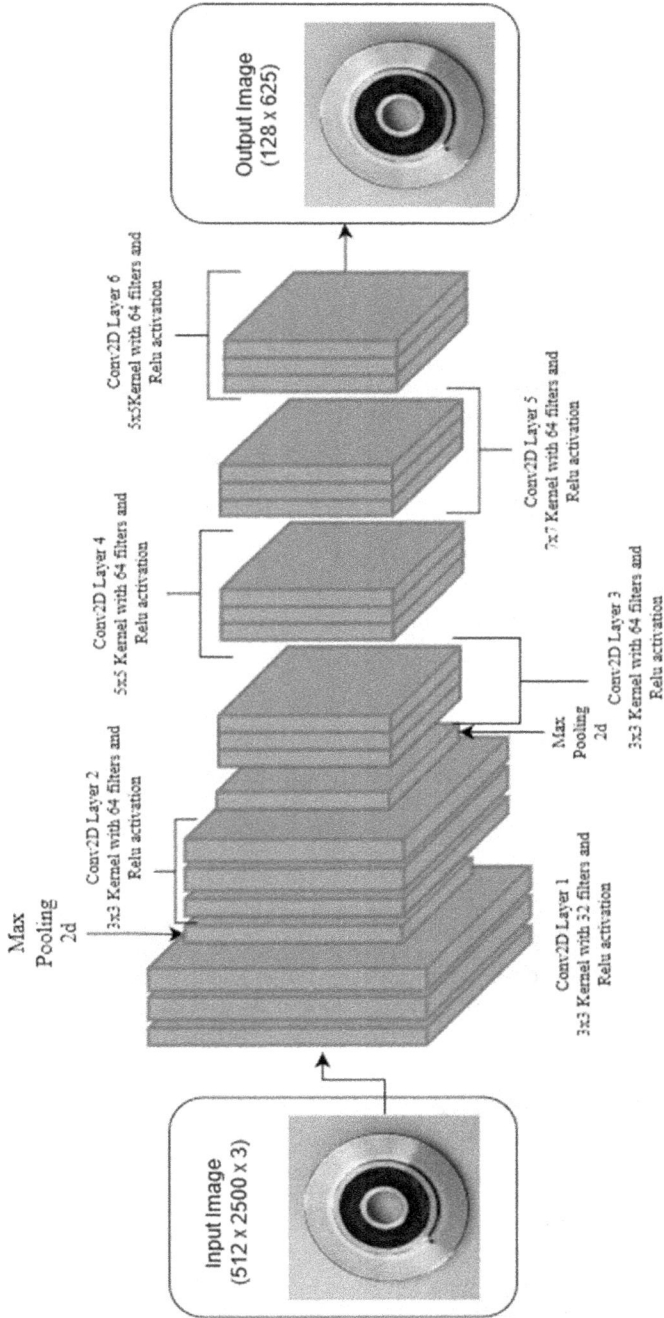

Figure 3.14 Segmentation network for defect detection.

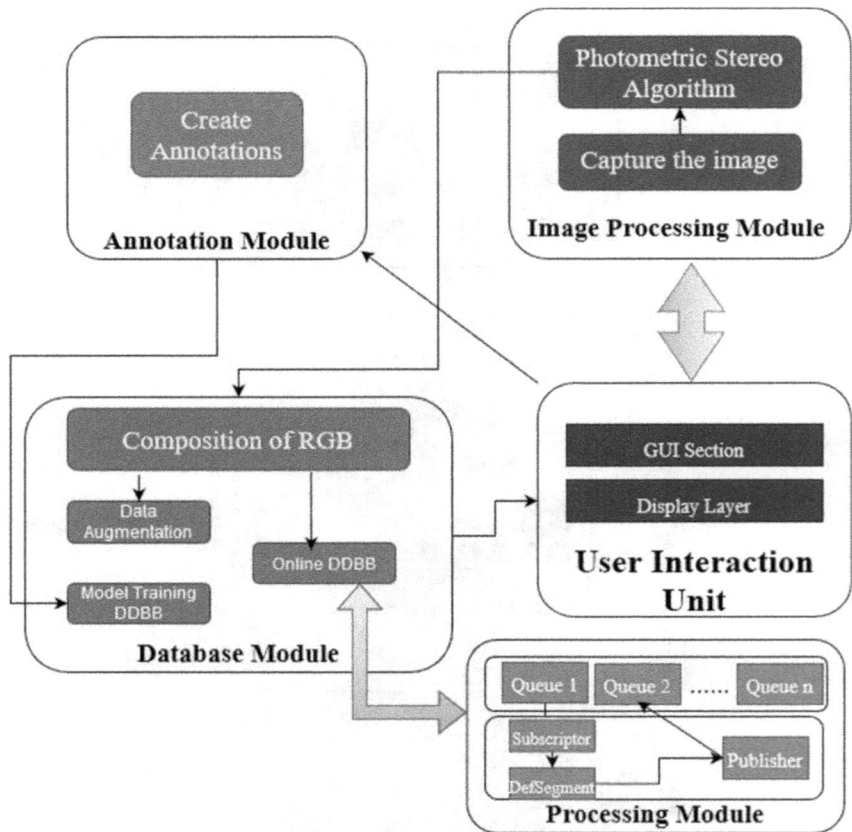

Figure 3.15 The system architecture of photometric stereo with segmentation network.

3.2.1.3 Motif discovery with CNN

The central concept behind this motif discovery is a time series analysis of large datasets—the process of achieving the image datasets in time series with the help of two steps. (i) Conversion of images into time series; (ii) Time series input used to determine the motifs. The algorithm process is deliberated in the following Figure 3.16.

3.2.1.4 EfficientNet-B0 with CNN

This approach assists with expanding the exactness at negligible computational expense. It is utilized for another scaling cycle in brain networks called compound scaling. These cycles are carried out at the pre-processing stage. Then, at that point, the pre-prepared pictures are brought into an essential convolutional brain organization. The EfficientNet-B0 standard

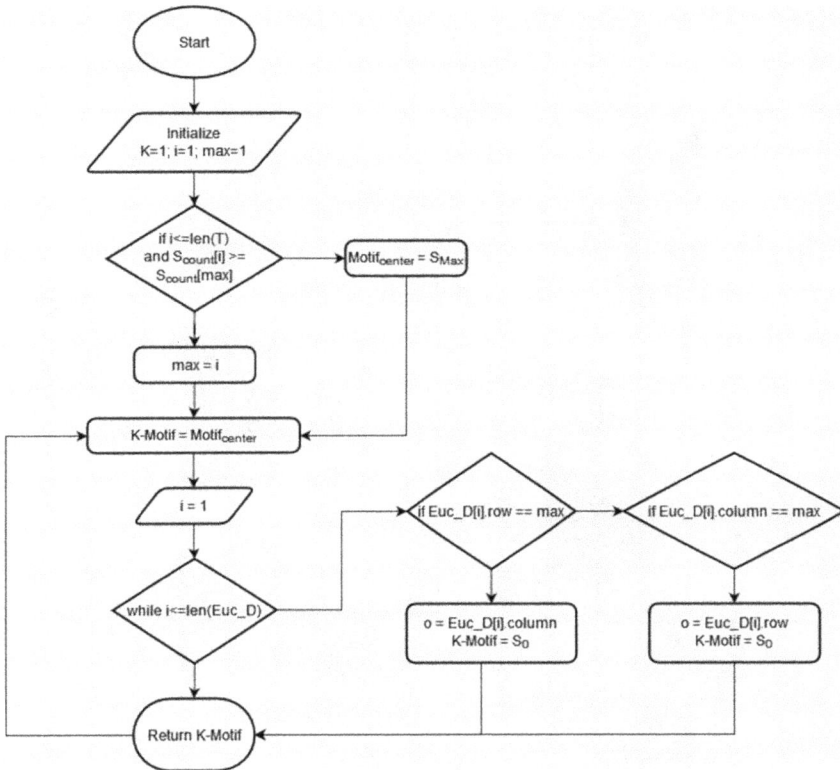

Figure 3.16 Motif algorithm process.

organization is the versatile gauge to assess convolutional networks' scaling strategy [8]. Figure 3.17 shows the process flow of Intelligent Machine Vision with CNN.

3.2.1.5 Multi-optical image fusion (MOIF) with CNN

The profound learning model for the combination picture information further develops the low-identification execution because of the uneven surface of a projecting item, and an optical recognition framework that integrates the photometric sound system procedure is utilized. Here, single pictures addressing an item's reflectivity, harshness, and slant are obtained using the optical framework and joined with a red, green blue (RGB) three-channel. Figure 3.18 shows the architecture of MOIF with CNN [9].

3.2.1.6 BoDoC methodology with support vector machine

The BoDoC methodology consists of two levels. The primary-level examinations of the information highlight utilizing a parallel characterization

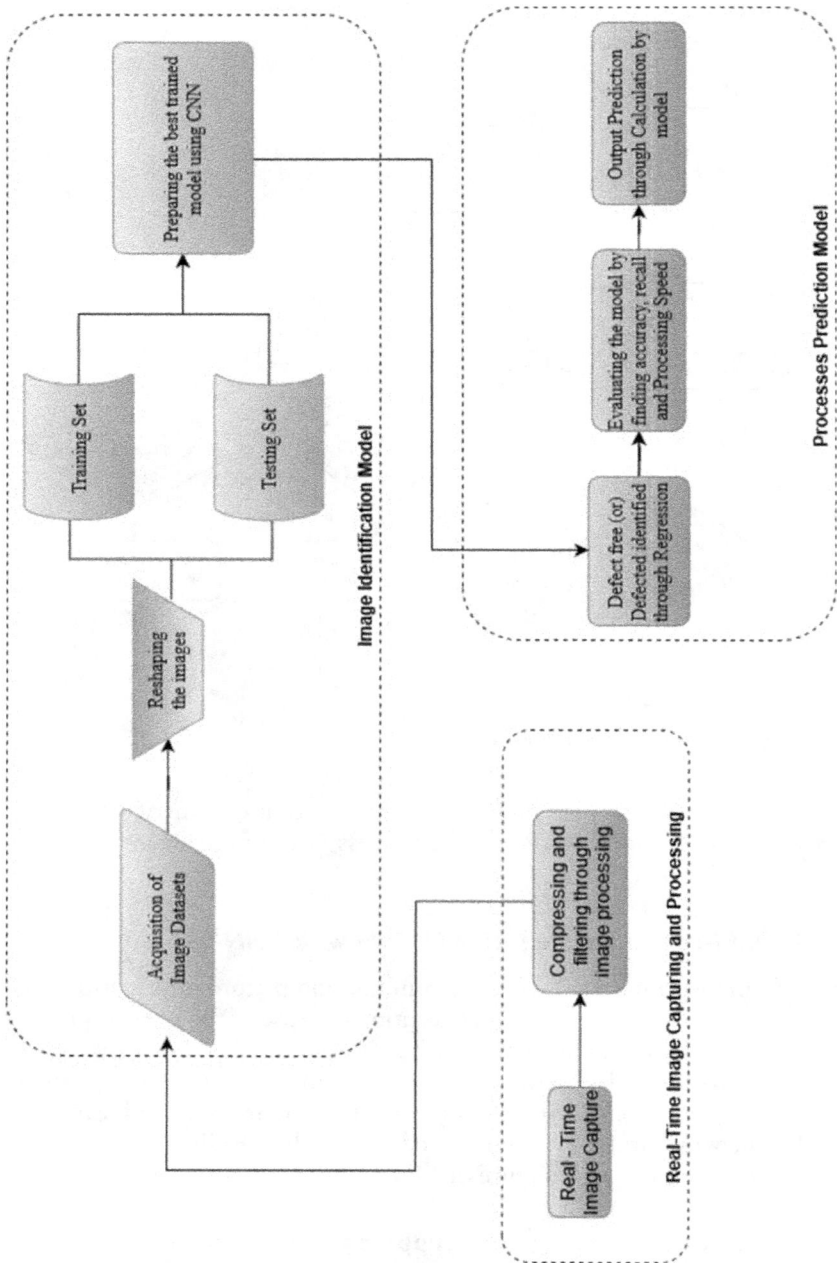

Figure 3.17 Process flow of intelligent machine vision with CNN system.

CNN Model to Predict and Classify the type of defect

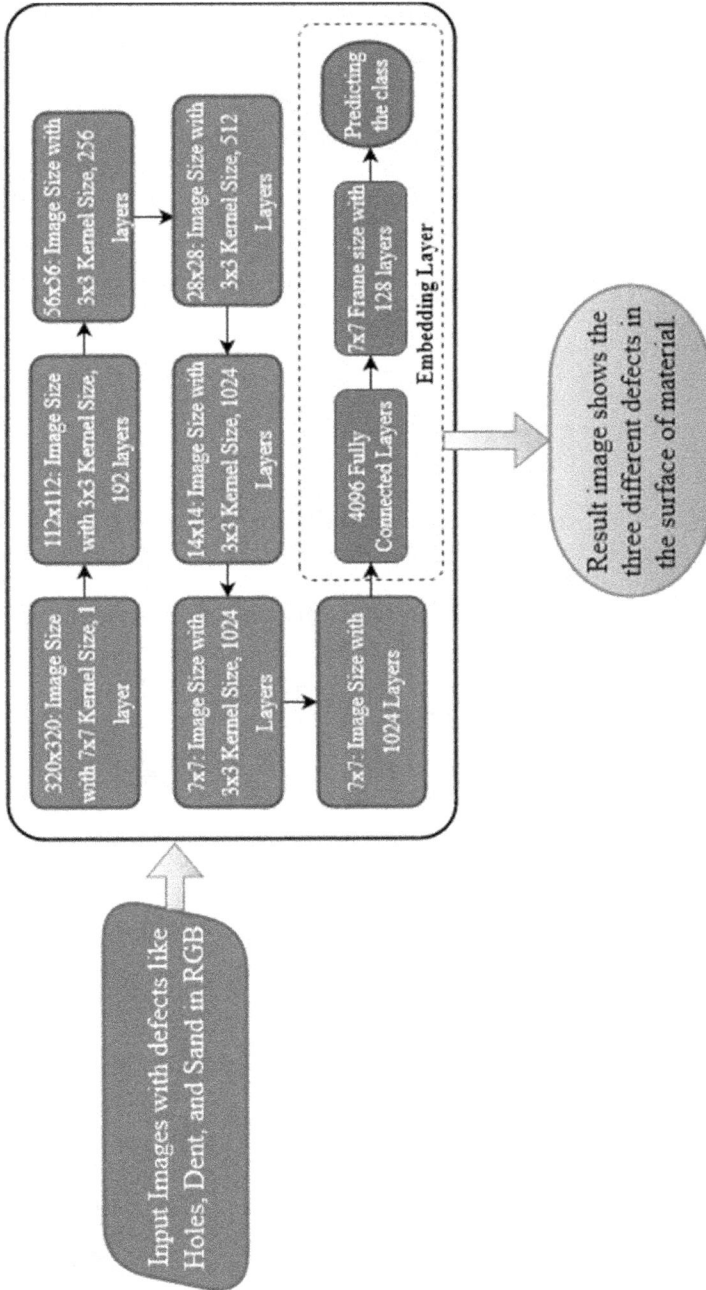

Figure 3.18 The architecture of MOIF with CNN.

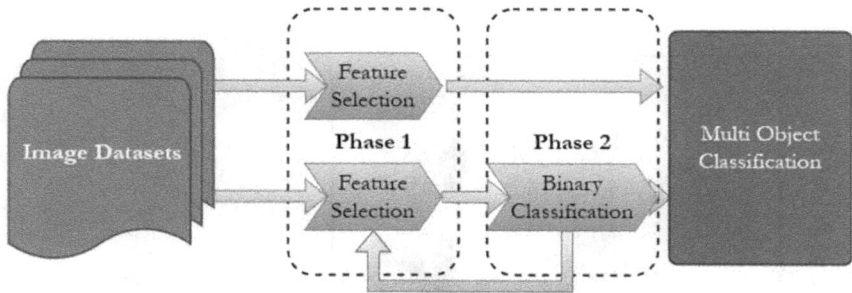

Figure 3.19 BoDoC methodology.

(or an assortment of them). The fundamental objective of this first stage is to make undeniable level portrayals of the data. Using this strategy, the less essential elements limit their effectiveness in the last characterization stage. Dissecting the exhibition of various calculations in this phase is attractive.

The subsequent stage characterizes the last multi-name order. To improve the limit of the order of the model, both the elements extricated initially, and also the result of the main stage are contributions to these subsequent AI models [11]. Figure 3.19 describes the primary and next steps in the BoDoC methodology.

3.2.1.7 SCN with ResNet-101 and Darknet-53

SCN is a deep CNN that can naturally remove undeniable level highlights from input pictures. It is therefore essential to involve the remaining module for SCN to stay away from the vanishing of the angle. The goal of significant level element maps is generally low, which leads to specific issues. The system is utilized in SCN to characterize significant level component maps at the pixel level, which decides if the pixel has a place with imperfection, so different deformities can be communicated on the element map [12]. Figure 3.20 shows the schematic process of SCN with ResNet.

3.2.1.8 Wasserstein generative adversarial nets (WGANs) with CNN

The Wasserstein generative adversarial net (WGAN) structure comprises two gaming organizations: a generator network which catches irregular commotion and produces counterfeit pictures, and a discriminator network which targets recognizing genuine and fake pictures [15]. The generator network is prepared to mislead the discriminator into seeing created pictures as genuine ones while the discriminator is prepared to separate the produced pictures. Figure 3.21 shows the systematic process of WGAN.

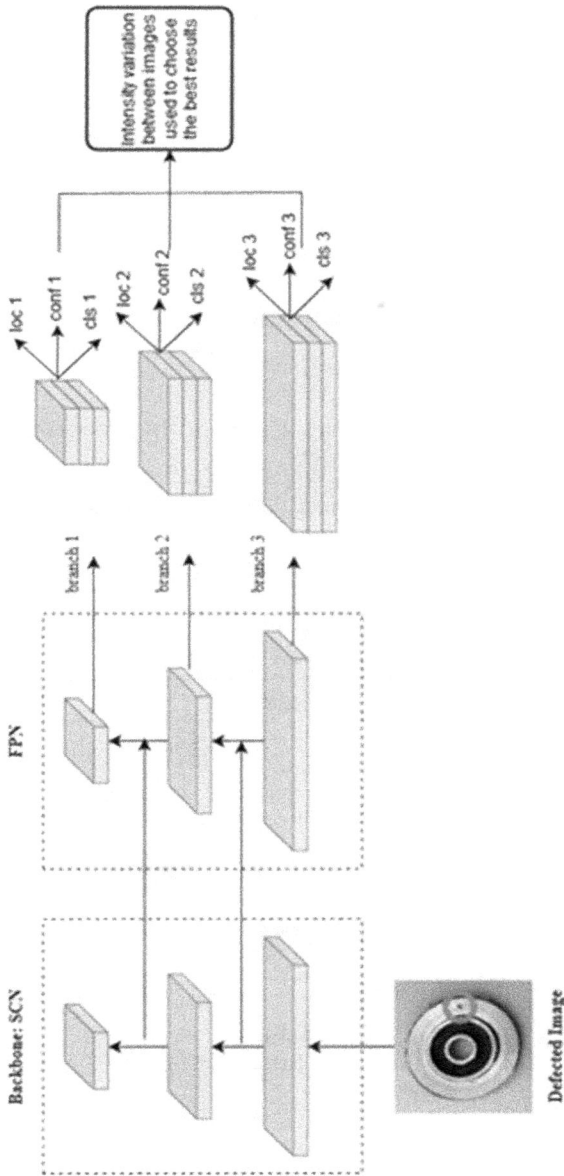

Figure 3.20 SCN with ResNet architecture schematic.

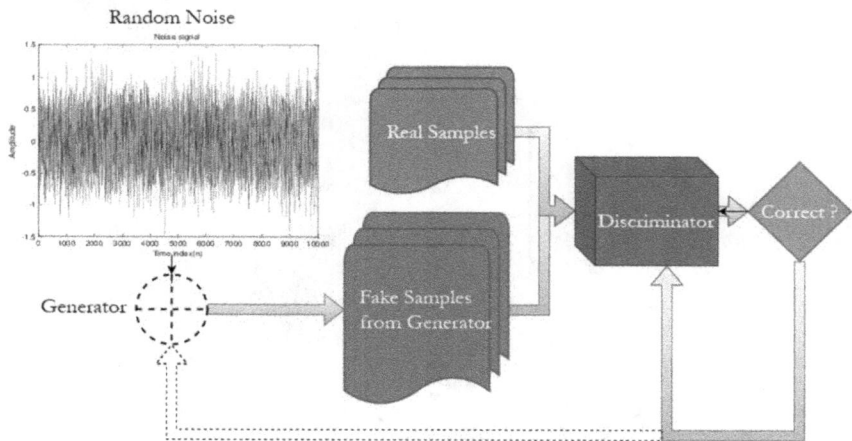

Figure 3.21 A systematic process of WGAN.

3.2.2 Discriminations of various approaches to defect detection

This section describes the various defect detection approaches in Table 3.2 and the results achieved through different techniques concerning the materials and methodologies in Table 3.3.

Table 3.2 Discriminations of different approaches

	Approaches				
Discrimination	*CNN with MVGG19*	*WGAN*	*Motif discovery with CNN*	*EfficientNetB0 with CNN*	*Segmented convolutional network with ResNet-101*
External source	Required	Partially required	Required	Required	Not required
Sample size to train the model	Significant (in 1000)	Small	Small	Small	Large
Type of input data	Labeled data used	Train model with small labeled data	Labeled data used	Pre-trained data	Unlabelled Data
Category of learning	Supervised Learning	Semi-supervised learning	Supervised Learning	Supervised Learning	Unsupervised Learning
Efficiency	High	Low	High	Moderate	Low

Table 3.3 Summary of different approaches and its result achieved

Methodology	Parameters measured	Result Achieved
CNN with MVGG19 Network [2]	Defect detection and object recognition in industries. (Casting, tools, Metal surface, Magnetic tile, Solar cell, Bridge crack)	6 Different industrial-related datasets defect detection and object recognition are done. (1) Casting - 97.88% (2) Defect - 77.62% (3) Magnetic - 92.67% (4) Tech - 94.23% (5) Bridge - 99.02% (6) Solar - 76.78%
Photometric stereo algorithm with customed segmentation network [18]	Material surface defects (NI- Nickel)	Accuracy of 95.60% is achieved for NI surface defects detection.
Visual Motif discovery approach [5]	Casting plate defects	Casting quality inspection is done through this article and achieved 97.14% accuracy, which is very high compared to CNN.
Bagging and Random Forest provide good predictability of defect identification in different point cloud densities of data. [16]	Plane surface and Barrel surface in turbine blades	Defects on the surface using Bagging and Random Forest algorithm for plane surface - 99.78% and barrel surface - 99.92% accuracy is achieved respectively.
EfficientNetB0 model trained in CNN for classification of product and Decision Tree algorithm for prediction process [8]	Casting plate defects	The classification model in this system achieved 96.88% accuracy, and the prediction model reached 99% of regression analysis.
Multi-optical Image Fusion (MOIF) with CNN. [9]	Casting product	The experimental results demonstrate that the model's mean Average Precision (mAP) was improved to 88%.
BoDoC methodology with Support Vector Machine (SVM: Pearson VII and Decision Tree (DT: Random Forest N = 100) for classification [11]	Casting at foundry	Results in real-world scenarios, with 91.305% precision.
SCN with ResNet-101 and Darknet-53 [12]	Surface of casting	The average accuracy of this system is 99.61% and 95.84%, respectively, which is significantly higher than the Faster-RCNN series and the YOLOv3 series.
Wasserstein Generative Adversarial Nets (WGANs) and feature extraction done through CNN [14]	Cast austenitic stainless steel (CASS)	The classification accuracy of MLP is 98.54% which is high compared to other classification methods like KNN and SVM.
CNN [10]	Casting product	Validated with 400 casting products and achieved 98% accuracy.

Table 3.4 Pros and cons

S. No.	Approach	Pros	Cons
1	Photometric stereo and Multi Optical Image Fusion (MOIF)	There is a compelling reason need to know the exact 3D connection between the test object and the camera or to utilize two cameras to catch 3D information	Impediments of distinguishing non-Lambertian surfaces such as polished metals
2	Stereoscopic vision	Appropriate for regions with enormous surface varieties and is exceptionally touchy to typical surface unsettling influences	It relies upon the inherent surface data of the object surface
3	Laser scanner	Duplicate the surface shape with the goal that it is non-contact, non-disastrous, and has high accuracy	The gear cost is high, and the computation sum is enormous
4	Structural light	High spatial goal and exactness	Complex computation and difficult to precisely align

3.2.3 Pros and cons of each approach

This section deliberates the different existing approaches' pros and cons in Table 3.4. Here, we discussed the processes related to the traditional way of inspecting the materials in industries.

3.3 SURVEY ON VARIOUS AVAILABLE EQUIPMENT FOR DEFECT DETECTION IN INDUSTRIES

This section describes the various equipment-based visual inspection processes in manufacturing. The visual inspection to detect the defect in an object can be done through the following methods: eddy current, ultrasonic, X-ray, Machine Vision, Magnetic powder and Osmosis testing [32] equipment available in industries.

- *Eddy current Testing:* This testing is done based on a magnetic field with an excitation coil applied in a time-varying pattern in the defective object and inspects the defected region by the conduction of magnetic field and defective materials. It is suitable to detect corrosion and material stress defects on the surface of an object.

- *Ultrasonic Testing:* Here, the ultrasonic waves are passed on the surface of the defected material. Echo/feedback of ultrasonic waves from the material surface identifies the defective region and its depth on the material's surface. It is mainly used to find defects in wood materials.
- *X-ray Testing:* This testing passes the X-ray on the material surface and tests the object based on the throwing back of rays in the targeted thing; it will show the defective region in the display, and we can visualize the defected part at a high-intensity level. Since it generates the radiation, it may cause a radiation effect on the user.
- *Machine Vision Detection:* This is a non-conductive test on the surface. We can take an image of the defective objects and highlight the region in the testing object with the assistance of artificial intelligence algorithms.
- *Magnetic Powder Testing:* We can visualize the defective region by applying the magnetic powder. This powder penetrates the surface of the object and shows the defective part is isolated.
- *Osmosis Testing:* It is reasonable to recognize surrenders in exceptionally permeable and non-permeable materials and enjoy specific upper hands over other general techniques. Nonetheless, the more significant part of the customer discovery strategies must depend on manual help to finish, which isn't exceptionally versatile and restricted by the hardware life and assembling exactness.

3.4 FUTURE RESEARCH CHALLENGES ON DEFECT DETECTION IN INDUSTRIES

In the future, producers are supposed to confront an undeniably questionable outer climate with a total impact of changes in client prerequisites, worldwide contests, and mechanical headway. Manufacturers face the test of further developing effectiveness and bringing down costs. Quality, adaptability, cost and time are among the primary methods for assembling organizations to get by. Quality Control (QC) procedures would be taken advantage of to assist associations with improving and developing their items and cycle to be acknowledged by clients [26]. Consequently, the execution of climate preservation, air-disposed assembling practices, and green innovation has all the earmarks of becoming prevalent.

The QC strategy will incorporate these ecological issues as its significant components. Straightforwardness and preparation for utilization would be the assumption for QC procedures to represent things to come. New research is in progress to plan a new QC approach to join the existing approaches and without a model framework to stay aware of the propelling innovation, extending fabricating interaction and developing item assortments. As a result of the increasing concern regarding supportable spots and assets for a group of people yet to come, makers should give additional thought to the ecological impact of their activities—the following highlighted process in the progress of challenges faced by the industrialists.

3.4.1 Small object defect detection

The plan is to foster a 3D motion filter concerning some most recent profound learning approaches for precisely recognizing small items [6]. In deep learning, implementing newer technologies to detect the defect in smaller objects through the use of Augmented Reality (AR) [13] is another potentially promising future approach. The current situation in ventures to decide the deformities in the small item require a high-resolution image capture device. Henceforth, this framework will be more expensive. This framework is not viable with each outer environment; it expects space to assess the defected objects. For instance, unique lighting conditions surrender location in things are unbelievable with this framework. Consequently, there is a need to foster the framework regardless of any outside environment.

3.4.2 3D reconstruction of defected objects

The main challenge of determining the defect in the thing through 3D is fixing the point to construct the 3D image of an object [7]. We believe that one of the notable forms of visual acknowledgment is the recognition of surrender in the outer layer of the article and the guessing of the associated 3D model to determine the abandoned form in the 3D layer [32]. This current situation is similar to what we would typically imagine as a 3D model of an entire table when we look at one side of a table. To determine the 3D level table of surface imperfections and imperfections, patterns can be remade and mixed using different techniques, although such methods depend on the collection. They are of the same class as a 3D-shaped peak. Finishing 3D models of different types of articles in complex situations requires more attention.

3.4.3 Depth analysis of the defected region

The depth map can be achieved through using the color image as reference, and feature points [19] are performed to calculate the depth of the defective part in the form of a 3D view to give an accurate measurement of the dimensions. This task is the most challenging for the industrialists. The laborers in the assembling enterprises deal with extreme issues to break down the profundity of the abandoned area in the article precisely. In the current venture, the producers utilize different strategies such as a photometric sound system, an X-beam, magnetic powder infiltrations, and so forth [32], as referenced above in this chapter. With these techniques, the precision of the profundity in the absconded district is not feasible and gives unseemly outcomes in some cases. Depth analysis shows the industrialists a significant scope in investigating the deepness of the imperfection in the items.

3.5 CONCLUSION

This chapter exposes various upcoming technologies for the inspection of objects such as AR-based inspection and preparing an ensemble neural network to classify the defected objects. Continuous expansion is expected to accomplish expanded exactness. Increasing amounts of continuous information is required for training and testing to develop the best exertion responses for the company groups dealing with deformity and non-imperfection. The following are the essential characteristics to determine the flaws in the industrial objects,

> ➢ Non-horrendous imperfection recognition strategies ought to be coordinated to acknowledge multi-modular deformity discovery of assembling items, which can have expansive application possibilities in the field of deformity identification.
> ➢ In the actual creation process, the imperfection data of the item is not just shown on the outer layer of the fabricated item. It additionally requires the utilization of 3D deformity discovery techniques to distinguish the 3D surface qualities of the test.
> ➢ High accuracy identifiable proof novelty. During the time spent preserving the image, the apparent attributes of the item change significantly with different lighting conditions and shooting points and distances. Because of the other bases of the identifying object, multiple buzzing and incomplete barriers of the announced example can significantly affect the detection results.
> ➢ Instructions to streamline the nature of picture procurement, work on the precision of the applicant box, separate elements thoroughly and precisely for learning, and concentrate highlights of small size targets.
> ➢ After recognizing the faulty products, the employees can assist in the opportune killing of the defective ones if the imperfection detection framework is combined with the early advance notice framework to provide timely advance notice. Or, on the other hand, with the arranging situation, the controller to kill the damaged items, furthermore, can likewise lay out a recognizability framework to check the creation interaction will make the thing surrenders steps, and convenient advancement of the creation cycle, to decrease the creation cost.

Hence, we hope this chapter will assist modern enterprises and scientists in understanding the exploration progress of item imperfection location innovation in profound learning and conventional deformity identification.

REFERENCES

[1] Lerui Chen, Jianfu Cao, Kui Wu, Zerui Zhang. (2022). Application of Generalized Frequency Response Functions and Improved Convolutional Neural Network to Fault Diagnosis of Heavy-duty Industrial Robot. *Robotics and Computer–Integrated Manufacturing*, 73, p. 102228.

[2] Ioannis D. Apostolopoulos, Mpesiana A. Tzani. (2022). Industrial Object and Defect-Recognition Utilizing Multilevel Feature Extraction from Industrial Scenes with Deep Learning Approach. *Journal of Ambient Intelligence and Humanized Computing*, pp. 1–14, doi: 10.1007/s12652-021-03688-7.

[3] Shikun Chen, Tim Kaufmann. (2022). Development of Data-Driven Machine Learning Models for the Prediction of Casting Surface Defects. *Metals*, 12, p. 1.

[4] Batta Mahesh. (2020). Machine Learning Algorithms – Review. *International Journal of Science and Research (IJSR)*, 9(1), pp. 381–386. ISSN: 2319-7604.

[5] Amanjeet Singh Bhatia, Rado Kotorov, Lianhua Chi. (2022). Casting Plate Defect Detection Using Motif Discovery with Minimal Model Training and Small Data Sets. *Journal of Intelligent Manufacturing*, 34, pp. 1731–1742.

[6] X. Gao, S. Ram, R.C. Philip, J.J. Rodríguez, J. Szep, S. Shao, P. Satam, J. Pacheco, S. Hariri. (2022). Selecting Post-Processing Schemes for Accurate Detection of Small Objects in Low-Resolution Wide-Area Aerial Imagery. *Remote Sensing*, 14(2), p. 255.

[7] A. Sungheetha. (2021). 3D Image Processing using Machine Learning based Input Processing for Man-Machine Interaction. *Journal of Innovative Image Processing*, 3(1), pp. 1–6.

[8] Tajeddine Benbarrad, Marouane Salhaoui, Soukaina Bakhat Kenitar, Mounir Arioua. (2021). Intelligent Machine Vision Model for Defective Product Inspection Based on Machine Learning. *Journal of Sensor and Actuator Networks*, 10(1), pp. 1–18.

[9] Jong Hyuk Lee, Byeong Hak Kim, Min Young Kim. (2021). Machine Learning-Based Automatic Optical Inspection System with Multimodal Optical Image Fusion Network. *International Journal of Control, Automation and Systems*, 19(10), pp. 3503–3510.

[10] Thong Phi Nguyen, Seungho Choi, Sung-Jun Park, Sung Hyuk Park, Jonghun Yoon. (2021). Inspecting Method for Defective Casting Products with Convolutional Neural Network (CNN). *International Journal of Precision Engineering and Manufacturing-Green Technology*, 8, pp. 583–594.

[11] Iker Pastor-López, Borja Sanz, Alberto Tellaeche, Giuseppe Psaila, José Gaviria de la Puerta, Pablo G. Bringas. (2021). Quality Assessment Methodology Based on Machine Learning with Small Datasets: Industrial Castings Defects. *Neurocomputing*, 456, pp. 622–628.

[12] Junjie Xing, Minping Jia. (2021). A Convolutional Neural Network-Based Method for Workpiece Surface Defect Detection. *Measurement*, 176, p. 109185.

[13] E. Marino, L. Barbieri, B. Colacino, A.K. Fleri, F. Bruno. (2021). An Augmented Reality Inspection Tool to Support Workers in Industry 4.0 Environments. *Computers in Industry*, 127, p. 103412.

[14] Jin-Gyum Kim, Changheui Jang, Sung-Sik Kang. (2022). Classification of Ultrasonic Signals of Thermally Aged Cast Austenitic Stainless Steel (CASS) Using Machine Learning (ML) Models. *Nuclear Engineering and Technology*, 54, pp. 1167–1174.

[15] Xinyi Le, Junhui Mei, Haodong Zhang, Boyu Zhou, Juntong Xi. (2020). A Learning-Based Approach for Surface Defect Detection Using Small Image Datasets. *Neurocomputing*, 408, pp. 112–120.

[16] Rui Li, Mingzhou Jin, Vincent C. Paquit. (2021). Geometrical Defect Detection for Additive Manufacturing with Machine Learning Models. *Materials & Design*, 206, p. 109276.

[17] Peng Wang, Yiran Yang, Narges Shayesteh Moghaddam. (2022). Process Modeling in Laser Powder Bed Fusion towards Defect Detection and Quality Control via Machine Learning: The State-of-the-Art and Research Challenges. *Journal of Manufacturing Processes*, 73, pp. 961–984.

[18] Fátima A. Saiz, Iñigo Barandiaran, Ander Arbelaiz, Manuel Graña. (2022). Photometric Stereo-Based Defect Detection System for Steel Components Manufacturing Using a Deep Segmentation Network. *Sensors*, 22, p. 882.

[19] L. Zhai, D. Chen. (2020). Image Real-Time Augmented Reality Technology Based on Spatial Color and Depth Consistency. *Journal of Real-Time Image Processing*, 18(2), pp. 369–377.

[20] R. Kayalvizhi, S. Malarvizhi, Siddhartha Dhar Choudhury, Anita Topkar, P. Vijayakumar. (2020). Detection of Sharp Objects using Deep Neural Network-based Object Detection Algorithm. In *4th International Conference on Computer, Communication and Signal Processing (ICCCSP)*. IEEE.

[21] Max Ferguson, Ronay Ak, Yung-Tsun Tina Lee, Kincho H. Law. (2017). Automatic Localization of Casting Defects with Convolutional Neural Networks. *IEEE International Conference on Big Data*, 1726–1735. ISBN: 978-1-5386-2715-0.

[22] Xiaoxin Fang, Qiwu Luo, Bingxing Zhou, Congcong Li, Lu Tian. (2020). Research Progress of Automated Visual Surface Defect Detection for Industrial Metal Planar Materials. *Sensors*, 20, p. 5136.

[23] Cheng Jin, Xianguang Kong, Jiantao Chang, Han Cheng, Xiaojia Liu. (2020). Internal Crack Detection of Castings: A Study Based on Relief Algorithm and Adaboost-SVM. *The International Journal of Advanced Manufacturing Technology*, 108, pp. 3313–3322.

[24] Vijayakumar Ponnusamy, Diwakar R. Marur, Deepa Dhanaskodi, Thangavel Palaniappan. Deep Learning-Based X-Ray Baggage Hazardous Object Detection–An FPGA Implementation. *Revue d'Intelligence Artificielle*, 35(5), pp. 431–435.

[25] P. Stavropoulos, P. Foteinopoulos. (2018). Modeling of Additive Manufacturing Processes: A Review and Classification. *Manufacturing Review*, 5, p. 2. doi: 10.1051/mfreview/2017014.

[26] H.M. Judi, R. Jenal, D. Genasan. (2011). *Quality Control Implementation in Manufacturing Companies: Motivating Factors and Challenges*. IntechOpen. Available at: https://www.intechopen.com/chapters/14857

[27] T. Czimmermann, G. Ciuti, M. Milazzo, M. Chiurazzi, S. Roccella, C.M. Oddo, P. Dario. (2020). Visual-Based Defect Detection and Classification Approaches for Industrial Applications—A Survey. *Sensors*, 20(5), p. 1459. doi: 10.3390/s20051459.

[28] A.J. Chittilappilly, K. Subramaniam. (2017). SVM-Based Defect Detection for Industrial Applications. In *2017 4th International Conference on Advanced Computing and Communication Systems (ICACCS)*. doi: 10.1109/ICACCS. 2017.8014696.

[29] I. Ioniţă, D. Şchiopu (2010). Using Principal Component Analysis in Loan Granting. *Seria Matematică -Informatică -Fizică*, LXII(1), pp. 88–96.

[30] R. Salakhutdinov. (2015). Learning Deep Generative Models. *Annual Review of Statistics and Its Application*, 2(1), pp. 361–385. doi: 10.1146/annurev-statistics-010814-020120.

[31] M. del Rosario Martinez-Blanco, V.H. Castañeda-Miranda, G. Ornelas-Vargas, H.A. Guerrero-Osuna, L.O. Solis-Sanchez, R. Castañeda-Miranda, J. María Celaya-Padilla, C.E. Galvan-Tejada, J.I. Galvan-Tejada, H.R. Vega-Carrillo, M. Martínez-Fierro, I. Garza-Veloz, J.M. Ortiz-Rodriguez. (2016). *Generalized Regression Neural Networks with Application in Neutron Spectrometry*, [Internet]. Artificial Neural Networks - Models and Applications. InTech, United Kingdom. doi: 10.5772/64047.

[32] J. Yang, S. Li, Z. Wang, H. Dong, J. Wang, S. Tang. (2020). Using Deep Learning to Detect Defects in Manufacturing: A Comprehensive Survey and Current Challenges. *Materials*, 13(24), p. 5755. doi: 10.3390/ma13245755.

Chapter 4

Manufacturing Data Performance Prediction and Optimization

A. Joshi
Panimalar Engineering College, Chennai, Tamil Nadu, India

Dahlia Sam and S. Sendilvelan
M.G.R. Educational and Research Institute, Maduravoyal, Tamil Nadu, India

K. Jayanthi
KGiSL Institute of Information Management, Coimbatore, Tamil Nadu, India

N. Kanya and N. Ethiraj
M.G.R. Educational and Research Institute, Maduravoyal, Tamil Nadu, India

CONTENTS

DOI: 10.1201/9781003257714-4

4.1 INTRODUCTION

Artificial intelligence (AI) is the modeling and simulation of intelligent behavior using computer programs. This concept considers the data structures that are used for the representation of knowledge, and the algorithms that are necessary to put that information to use, as well as the programming languages and methods that are utilized to implement those algorithms. It would be more accurate to say that the intellectual history of AI will serve as the primary focus of our introduction. The brilliant work of Aristotle should serve as this kind of history's natural launching point. Aristotle is credited with weaving together the wonders, insights, and Greek tradition's early fears with the thorough logical thought and investigation that would later establish the standard for contemporary science. In his book on "Metaphysics," Aristotle began by stating, "All men by nature seek to know." He then proceeded to develop a science out of things that do not change through time. This science included his cosmology and religion. In his *Physics*, he referred to the "study of things that change" as his "philosophy of nature". The explanation of human epistemology that Aristotle presents in his book *Logic*, which investigates how people "know" their reality, is, on the other hand, more relevant to the study of AI. In *Logic*, he investigated the question of whether or not certain statements may be considered to be "true" simply on the basis that they are linked to other things that are generally acknowledged to be the case. For example, given that Socrates is a man and that the saying goes that "all men are mortal," it is reasonable to deduce that "Socrates is mortal" based on these two pieces of evidence. This argument demonstrates what is referred by Aristotle to as a syllogism through the application of deductive modes of reasoning (logical reasoning when the conclusion is derived from two related premises). However, it took another two thousand years for the formal axiomatic reasoning to fully emerge, and this development occurred only in the twentieth century. Aristotle, Alan Turing, Alfred Tarski, and many other thinkers have all made a contribution to its development at various points in time. The current study of the mind and its organization is required to take into consideration the distinction between the human mind and the world around it, between conceptions about things and the objects themselves. The program compiled a large amount of data relevant to the phenomena; from that data, it was possible to infer several different physical principles. It's also interesting to note that the Turing Machine and proofs of computability was useful for the general-purpose algorithm to produce scientific proofs.

Industry 4.0 concepts of smart factories have been developed to conceive of a manufacturing paradigm technology-oriented that can achieve agility, flexibility, decentralization of production processes, and customization. This paradigm is what is known as a smart factory. In addition, Industry 4.0 necessitates a paradigm change away from the old concept of lean thinking, which is characterized by restricted adaptability, and toward an approach that is focused on production management. The needs of customers are currently being actively met by modifying manufacturing processes to deliver a wide variety of low-cost products. This is carried out to achieve the needs of customers. As a direct consequence, today's industrial environment is characterized by significantly greater product variability, shorter product life cycles, and increased levels of global competition. The ability to plan and control production is an essential component of any industrial system, production planning and control (PPC). Because of this aspect, the decision-making procedures regarding production, delivery and profitability are likely to be drawn out. On the other hand, the data obtained during the manufacturing life cycle could prove to be beneficial in applying machine learning techniques for cost reduction and simultaneously improving quality, productivity, and sustainability. This concept is backed up by an increasing body of theoretical research that offers fresh machine learning and artificial intelligence (ML/AI) methods for production planning and control. Despite this, one of the most significant barriers to the implementation of ML and AI in PPC is a lack of expertise in the myriad of application domains and the scopes connected with them.

An increasing number of programs have also started to work on the synthesis side of things, demonstrating the relevance of intelligence programs in the domains of science and engineering beyond very intricate analysis. Computer-controlled robots that should be employed in agriculture to eradicate pests, prune trees, and selectively pick mixed crops are some applications with a long-term focus. In the manufacturing industry, dangerous and monotonous tasks such as assembly, inspection, and maintenance should be carried out by computer-controlled robots. Computers also provide clinicians with assistance in diagnosing patients, monitoring the states of their patients, and administering treatment. In the realm of housework, computers should be able to offer guidance on topics such as cooking and grocery shopping, as well as making labor-saving contributions such as mopping floors, doing laundry, and conducting other maintenance tasks. It's a common fallacy that artificial intelligence must be used for business purposes in order to save money by replacing human workers with robots. However, in the corporate world, people are more concerned with increasing their prospects than with reducing expenses. Because bright people and intelligent computers have complementary skills, they can take advantage of possibilities that neither could exploit separately. Furthermore, since we do not yet know how to give computers all of the observation, thinking, and active talents that people possess, the total replacement of a human worker is both challenging and unattainable. In the realm of business, computers can be of assistance in the process of locating relevant information, scheduling labor, allocating resources, and identifying noteworthy regularities in databases. Similarly, in the field of engineering, computers can be of assistance in the development of more efficient control techniques, the creation of the ideal design, the explanation of previous judgments, and the identification of potential dangers in the future.

Instead of focusing on AI research and algorithms mostly utilized logical approaches. In addition, researchers in computer science and artificial intelligence have moved their focus away from working on neural network projects. The two fields of artificial intelligence and machine learning become irreconcilably opposed as a result of this. Up until that point, artificial intelligence has been trained using a program referred to as machine learning. Machine learning for the industry, which consisted of a significant number of researchers and technicians, was restructured into a separate discipline and battled for almost ten years after the separation.

The primary objective of the sector has switched from the software development of artificial intelligence to the actual problem solving of improving service delivery. It has switched its focus away from the methodologies of AI research and strategies utilized in probability theory and statistics. The ML industry continues to place a primary emphasis on neural networks throughout this period and then experienced explosive growth. The majority of this success can be attributed to the expansion of the Internet, which has allowed businesses to capitalize on the availability of digital data increase and the increased capacity to distribute their services through the medium of the Internet.

To maximize the potential of technologies related to artificial intelligence, ML is an essential component. ML is frequently confused with artificial intelligence (AI) due to its capacity for learning and making decisions; however, in reality, ML is a subfield of AI. It was a component of the development of AI up until the late 1970s. After that, it went its separate way and developed in its own right. Learning by machine has emerged as a crucial problem-solving method for cloud computing and online commerce, and it is currently being implemented in a diverse range of cutting-edge technologies. The primary objective of the sector switched from artificial intelligence software to the development of actual problem-solving of improving service delivery.

Today, for many firms, ML is an important component of research and business. It helps computer systems perform better over time by using algorithms and neural network models. As demonstrated in Figure 4.1, the mathematical model created by the machine learning algorithms utilizes sample data, sometimes referred to as training data, without being specifically programmed to make judgments.

4.2 PRODUCTION PLANNING, AND CONTROL (PPC)

Production planning and control (PPC) must be concerned with how the plans are carried out, including the precise scheduling of tasks, the distribution of workloads among machines (and workers), and the observation of

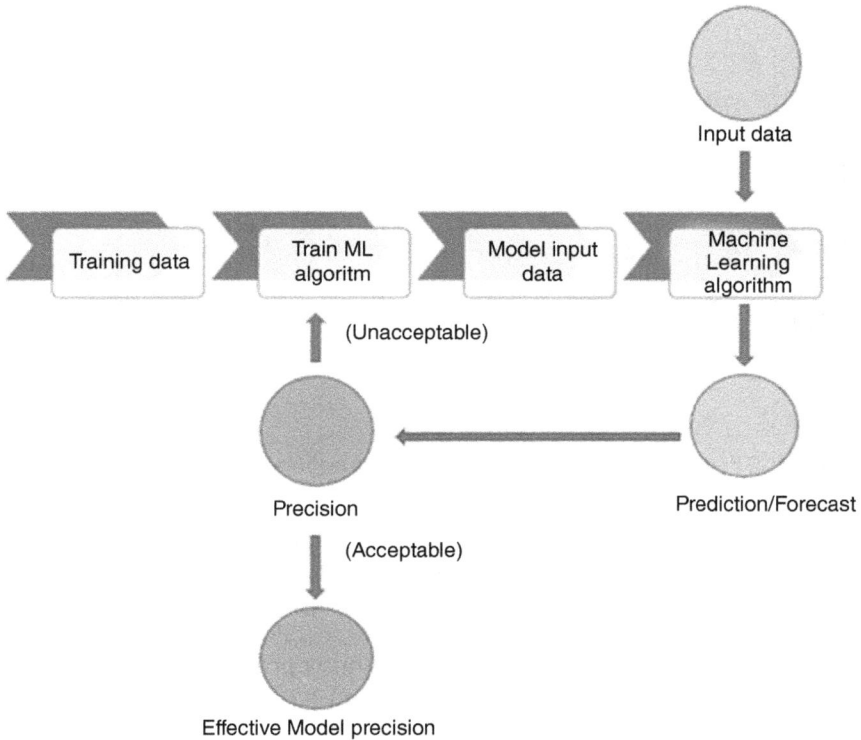

Figure 4.1 Machine learning methodologies.

the real-time movement of the system. Production is the process of transforming raw materials converted into finished items. The act of producing is a planned action. The production process requires the most effective use of all the natural resources that are at hand, including labor, capital, equipment, materials, and time. Organizing and managing production, depending on the type of business, you may need to communicate and work with even more divisions, such as marketing, production, warehouse, and logistics. The numerous marketing departments provide the PPC departments with the information they need to plan and regulate production. A production plan is established in production planning and control using information from both marketing and production. The production plan offers a distinct picture of how the available resources for manufacturing will be utilized in the production process. The production department receives the production plan that has been prepared for them. This department is responsible for making the products following the strategy. Production planning and management should ultimately be geared toward making a positive contribution to the bottom line of the business.

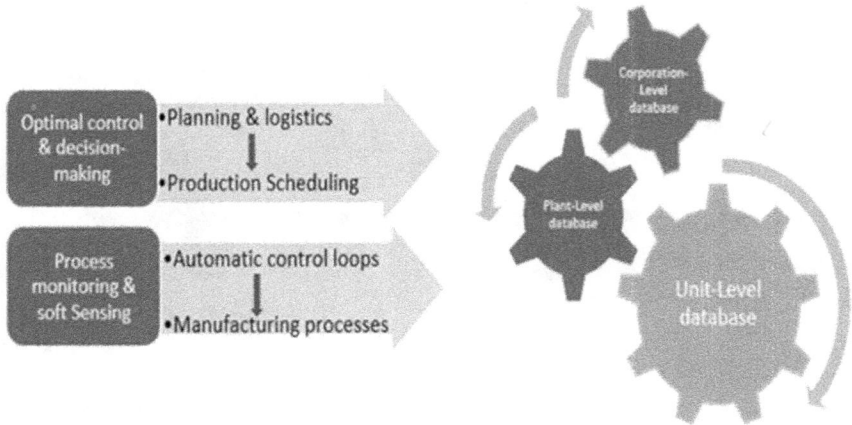

Figure 4.2 Production, planning, and control of machine learning.

The PPC is a very important decision that must unavoidably be made to guarantee an effective and cost-effective output. The ability to produce goods following plans is an essential component of any industrial sector. The acronym production planning and control (PPC) refers to a tool that is used to organize the entirety of a manufacturing process of production systems. This is done before the actual production activities begin. PPC involves the organization and planning of the production process. In the main, it encompasses the complete organization and all its aspects, including quantity, quality, delivery schedule, and production cost.

To provide the required manufacturing results in terms of quantity, time, quality, and location, it is ultimately necessary to control the supply and plan movement of labour, materials, and other production-related operations. This offers a structural framework and a prescribed guidelines for the productive transformation of labour, materials, and other necessary inputs into a finished good, as depicted in Figure 4.2.

Any company that implements a production system that is based on PPC will experience several benefits for a variety of different functional operations. These benefits include the following items:

a) By avoiding the rush at the last minute, production control can cut down on the number of emergency orders and overtime tasks that plants have to perform, hence lowering their overhead costs.
b) Problem areas of bottleneck get reduced: Because production control maintains a steady line and flow of work, there is no accumulation of unfinished work or work that is in the process of being completed.
c) Reduced costs: Proper production management optimizes the use of people and machines, which keeps in-process inventories at an acceptable level, improves raw material inventory management, lowers

storage and material handling costs, helps in maintaining quality, and reduces rejects, all of which contribute to a decrease in the production unit cost.

d) Efficient utilization of resources: This cuts down on the amount of time lost by workers as they wait for materials and ensures the most effective use of the equipment they have.

e) PPC is responsible for better coordination of the activities of the factory, which ultimately results in a more controlled and united effort on the part of the employees.

f) The increased productivity and efficiency brought about by PPC translate into fair pay, employment, job security, improved working conditions and job satisfaction, and excellent morale.

g) PPC guarantees production follows time schedules, which results in deliveries being made in accordance with the committed schedules, improving services for customers. As a result, the quality of services offered to clients improves.

4.2.1 Scope of PPC

Nature of the Inputs: The nature of the inputs that are used has a direct impact on the quality of the final product. Therefore, planning is performed to determine the nature of the many different sorts of inputs, which is a difficult procedure.

Input Quantity: The only way to prepare a product is to have an estimate of the required composition of the quantity of the inputs that are needed.

Coordination: This guarantees that the workforce, machines, and equipment all have the right amount of coordination with one another. This results in the prevention of waste as well as the uninterrupted flow of production.

A Better Handle on Things: Planning is required to have a better handle on things. When this occurs, and only then, is it possible to calculate the variances that lead to control of the production to compare the performance. A consistent supply of raw materials and other components can be ensured by careful planning of the materials, which also helps to prevent interruptions in production. The unbroken flow of manufacturing is made possible by the consistent delivery of raw materials and other suppliers. Utilization of Capacity: There is a pressing need to make efficient use of all of the resources that are now available. It is useful in reducing various expenses associated with production thanks to its contribution.

When there is proper production planning and management finished products can be promptly sent to the market at the right time. In addition, doing this fosters improved rapport with the clients. The planning and management

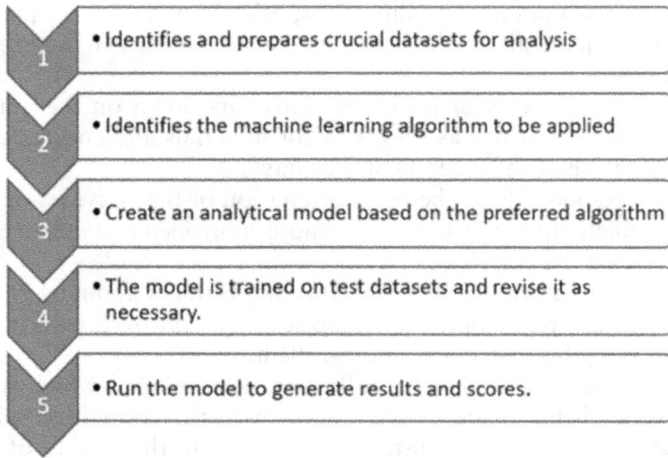

Figure 4.3 Machine learning process.

of the manufacturing process are sometimes compared to the nervous system of the industrial activity. The production activities that transform raw materials into completed goods or component parts are planned, coordinated, and managed by this function in the most efficient way feasible. The objective of this job is to use people, facilities, and material resources in any activity as efficiently as possible. The production activities are planned, coordinated, and controlled to achieve this, as indicated in Figure 4.3.

4.3 DYNAMIC SCHEDULING

The bulk of production systems works in dynamic environments, which implies that real-time occurrences that are typically inescapable and unforeseen may cause changes to the scheduled plans. Additionally, once a timetable is made available to the shop floor, it may no longer be viable. Such real-time occurrences include, for instance, the arrival of urgent jobs, machine issues, altered due dates, and so on. MacCarthy and Liu [1] addressed the nature of the gap between the scheduling theory and scheduling practice, the failure of classical scheduling theory to respond to the needs of practical environments, and recent trends in scheduling research which attempt to make it more relevant and applicable. When Cowling and Johansson [2] claimed that algorithms and scheduling models are unable to employ real-time data, they addressed a significant gap between scheduling practice and theory [2].

4.3.1 The dynamic scheduling problem

Numerous real-time events and their impacts on diverse industrial systems have been considered in the research on dynamic scheduling (Figure 4.4).

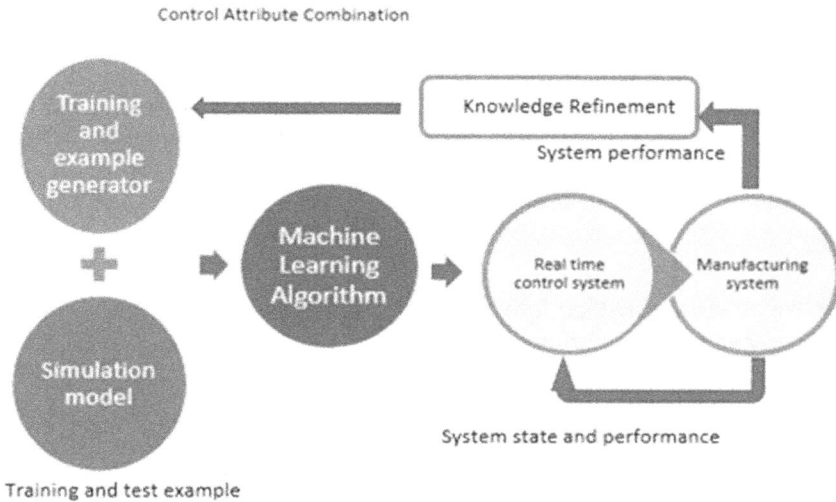

Figure 4.4 Dynamic scheduling.

Parallel machine systems, Single machine systems, job shops, flow shops and flexible manufacturing systems are among examples of these types of production processes. Two separate categories of real-time events have been established: (i) scheduling related to tasks, and (ii) scheduling related to resources. The three types of dynamic scheduling are: completely reactive scheduling, where decisions are made in real-time and dispatching rules are used; predictive-reactive scheduling, which is the most popular dynamic scheduling method used in manufacturing systems; and robust pro-active scheduling, which concentrate on creating reactive-predictive schedules. The effect of disturbance can be evaluated by calculating the amount of time that elapses between the actual completion of a project on the realized schedule and the time that was planned for its completion on the predictive schedule. The variation is brought under control by adding more time to the predictive timetable to achieve a high level of predictability.

4.3.2 Dynamic scheduling techniques

The issue of dynamic scheduling has been resolved using methods, including meta-heuristics, heuristics, neural networks, fuzzy logic, knowledge-based systems, hybrid methodologies, and multi-agent systems.

4.3.2.1 Heuristics

In the context of this discussion, heuristics refer to problem-specific schedule re-pairing strategies. These strategies do not ensure the discovery of an ideal schedule, but they do have the capacity to identify excellent solutions in a short amount of time. The repair strategy reschedules activities so that they will, at some point in the future, match up with the pre-schedule. Partial schedule repair reschedules only the operations in failure. The use of dispatching rules, which are essentially heuristics, has been an important factor in the development of entirely reactive scheduling. On the other hand, in a system of entirely reactive scheduling, there is no generation of a solid schedule in advance. Instead, real-time dispatching rules are used to select the next job that is waiting to be processed at a resource.

4.3.2.2 Meta-heuristics: Tabu search, simulated annealing and genetic algorithms

Meta-heuristics, such as simulated annealing, tabu search, and evolutionary algorithms, have all been successfully used in recent years to address problems with production schedules (Figure 4.5). Higher-level heuristics, known as meta-heuristics, steer local search heuristics away from the best options in the immediate area. Neighborhood search techniques built on the idea of exploring neighborhoods are known as heuristics for local searches. A local neighborhood search is a type of search that begins with a given solution and uses move operators to try repeatedly to move to a better solution within a properly specified neighborhood of the present solution. This type of search begins with a given solution. When there is no chance of finding a better solution in the vicinity of the existing solution, often known as the local optimum, the search process is finished. By creating good starting solutions and by tolerating inferior solutions or for the local search in a

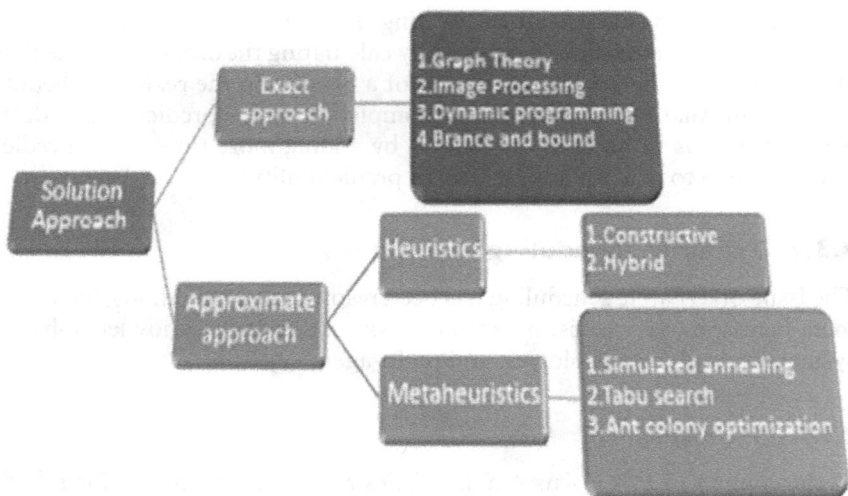

Figure 4.5 Meta-heuristics: Tabu search, simulated annealing and genetic algorithms.

more intelligent manner than just randomly producing early solutions, meta-heuristics such as simulated annealing, tabu search, and genetic algorithms improve to escape local optima by the local search. This approach is different from the conventional strategy of just giving out initial solutions at random.

In recent years static deterministic production scheduling issues have been successfully solved using simulated annealing, tabu search, and evolutionary algorithms in a number of settings, including open shops, job shops, flexible manufacturing systems, flow-hops, batch processing, and so on. However, in the context of flexible manufacturing systems the use of metaheuristics in dynamic scheduling has to date received relatively little research.

4.3.2.3 Multi-agent-based dynamic scheduling

Most industrial scheduling systems have typically been seen as a top-down, command-and-response process that largely utilize centralized and hierarchical concepts and have been created in industrial contexts, all of which contributes to this viewpoint. Both centralized and hierarchical scheduling systems rely primarily on central databases to ensure that the data they store is consistent throughout the entirety of the organization (Figure 4.6). Centrally, at the level of the supervisor, choices with regard to scheduling are made to optimize performance. These decisions are then transferred to the level of the manufacturing resource in order to be carried out. A centralized computer is given responsibility for scheduling, the distribution of resources, the monitoring of any deviations, and the distribution of corrective actions when a common architecture is used.

The presence of a central computer is the most significant disadvantage. This computer represents a hurdle that can restrict the store capacity, and also represents a single point of failure that has the potential to shut down the entire establishment. In addition, changing the configuration of a production system that is controlled hierarchically is a time-consuming and expensive process because it requires the rewriting of expensive software. The integration of various manufacturing system components results in a significant increase in the level of complexity of hierarchical scheduling systems. Another drawback is that the constant flow of information both upwards and down through the system can lengthen the amount of time it takes to make a decision. In addition, years of experience in the real world have shown that hierarchical centralized scheduling systems tend to struggle when it comes to reacting to disturbances; they may be unable to respond efficiently when there are real-time events present. When a disturbance occurs, the information is relayed to a higher level in the hierarchy. However, the new schedule does not begin to generate a new flow of commands that constitutes the reaction to the disturbance until after the scheduler has been modified to accommodate the information. This transfer of information both up and down results in a poor response time, which in turn leads to a low level of robustness. Although hierarchical and centralized scheduling

Start

Marker/POST
Ask to all AGV Agents to
send their Proposals

WAIT for Reponses
comingfrom
AGV agents

NO

PASS

If
reached?
NO

NO

If
reached?
YES

PASS

YES

YES

Pick the best
Proposal

Add Proposals
for Bidding

PASS

REPLY: Winner
Announcement

Stop

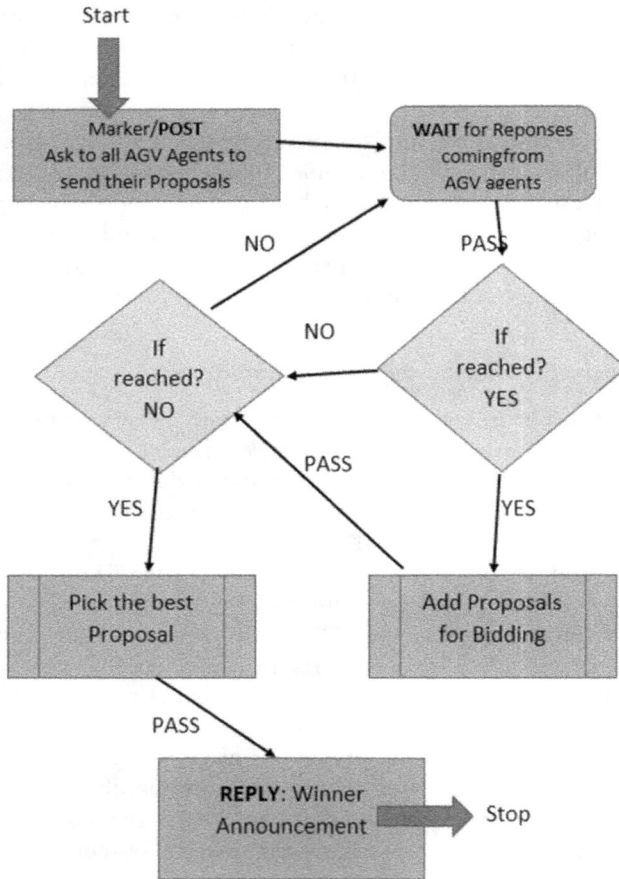

Figure 4.6 Multi-agent-based dynamic scheduling.

systems may produce globally superior schedules in environments in which real-time disruptions are uncommon, it is becoming increasingly clear that these kinds of scheduling approaches are not effective at all when it comes to reacting to highly dynamic environments. As a result, centralized and hierarchical scheduling is difficult to manage and also to reconfigure. It is also rigid, costly, and slow, making it impossible to satisfy the requirements of today's complex production environments.

4.4 PERFORMANCE EVALUATION AND MONITORING

At the beginning of the 20th century, mass production went from being a pipe dream to being a fully realized possibility. A subsequent transition from mechanization to digitalization has now taken place in the manufacturing industry. This has been regarded as a direct result of the development of Industry 4.0, a term that is used to describe the various processes and technologies that are used to modernize the manufacturing industry. This concept was coined in the year 2013. This has been applied to a wide range of different technologies: cloud computing, the Internet of Things (IoT), robotics, big data, artificial intelligence (AI), Blockchain, etc. These have led to a shift in how one functions. In addition, the term "Industry 4.0" refers to novel production methods that integrate contemporary forms of manufacturing technology, parts of manufacturing production, and human labor. In addition, the concept of Industry 4.0 identifies a number of new manufacturing patterns that may be used in any organization and which combine new forms of manufacturing technology, manufacturing parts, and personnel. It provides an alternative to the manufacturing process and produces an extraordinarily effective manufacturing system. These processes lower overall production costs while also improving product quality.

Due to the recent growth of the idea of Industry 4.0, the trending applications of AI/ML in the manufacturing industry and research necessitate a full analysis of the current state-of-the-art. This is necessary for the emerging applications of ML and AI in research and the manufacturing sector. In this section, we conduct a literature review on the application of AI and ML in industrial settings (Figure 4.7). In every production system, the monitoring and control of both the input and the output are extremely vital to the process of optimizing the system. The effectiveness of manufacturing

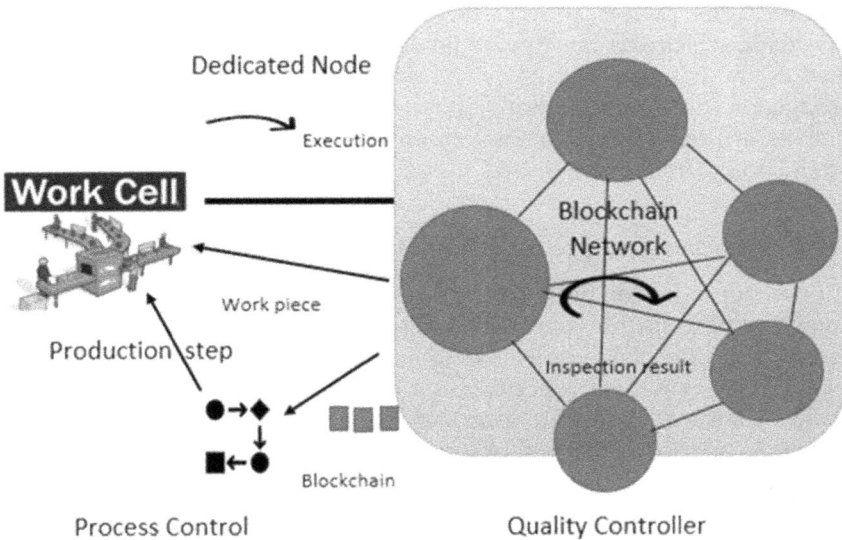

Figure 4.7 Performance evaluation and monitoring production cycle AI and blockchain.

systems encompasses a comprehensive range of technological and managerial endeavors. As stated previously, three categories can be used to categorize AI and ML applications in PPC-decision-intensive tasks: automated process control, monitoring, performance evaluation and dynamic scheduling. These three application categories mark a sizable step forward in Industry 4.0's efforts to enhance the automation of human-dependent processes, which sometimes need complex decision-making from production staff or management. For example, by recommending dynamic allocations for resources, projects, and production schedules, process rescheduling offers a greater degree of flexibility than traditional scheduling methods. Process rescheduling also gives more notable flexibility by suggesting dynamic assignments for activities, resources, and real-time production plans, to minimize disruptions to the production plans that were initially developed. The evaluation and monitoring of performance make it possible to conduct analytics of production performance in real-time and, as a result, propose operational modifications that could increase total production efficiency. The automation of processes makes it possible to automate production planning and control, and it also brings about considerable increases in the quality of work performed in occupations that require making difficult decisions during production planning.

When it comes to the production planning and management of industrial systems, one of the most important responsibilities is the monitoring and evaluation of production metrics. PPC provides for the discovery of bottlenecks, which assists with decision-making throughout the production processes. Bottlenecks can be found by monitoring and evaluating several manufacturing performances factors. The widespread application of industrial data, which has been made possible by developments in technology such as the IoT, smart sensors, and cloud manufacturing, has led to the production of a significant body of written work in this area. The following is a list of the important areas that call for monitoring in the context of production planning and control performance evaluation.

4.4.1 Performance prediction and optimization

Improving production metrics by utilizing historical production data in conjunction with industrial data is the quality emphasis of this research field. The detection of limits and bottlenecks in manufacturing is made possible using analytic real-time data collected from the processes and machines. These analyses consider the entire industrial ecosystem. In addition, the application of AI and ML talents allows for enhanced adaptability and flexibility, which is essential for achieving high levels of performance. Performance prediction, which is utilized in the context of PPC, is a significant application of AI and ML. To predict the deep drilling procedure success of under various cutting and cooling circumstances, Bustillo et al. [3] developed, with the use of regression trees and Principal Component Analysis (PCA),

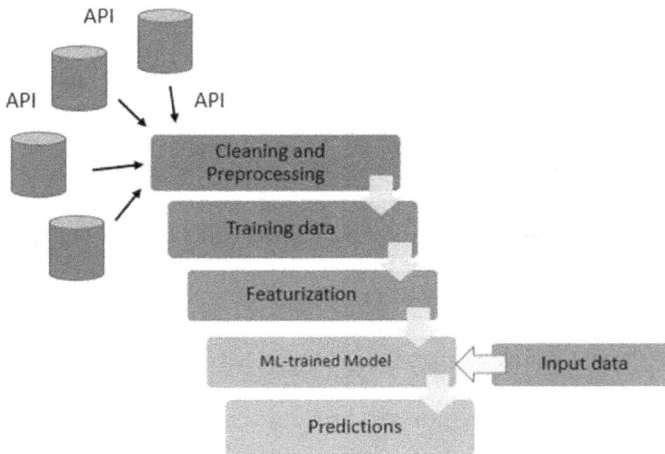

Figure 4.8 Performance prediction and optimization machine learning and artificial intelligence.

a hybrid approach that combines conditional inference trees with dimensionality reduction. This approach was a hybrid because it combined regression trees with conditional inference trees (PCA), as shown in Figure 4.8.

4.4.2 Process fault detection

The elimination of mistakes and the high-level maintenance of quality throughout the entire process are two of the most significant goals that can be pursued through monitoring and performance evaluation. This is one essential component of both the production planning and the management of the production. Our primary focus will be on the strategy that is based on ML and AI aims to enhance the overall performance of industrial systems via the application of these two technologies. It is possible to anticipate and prevent irregularities and failures in the manufacturing process. For instance, Morales et al. [4] suggested developing an early warning system for issues in egg production by employing a Support Vector Machine (SVM) approach [4].

4.4.3 Quality management and Quality 4.0

- At this point, the quality of the product, the service, and the process is essential for achieving long-term economic growth and keeping productivity high. The management and control of quality have attracted the attention of a significant number of academics and executives in business, and it is an essential subject for study and investigation.

The idea of "Quality 4.0" needs to be adopted by manufacturers to successfully incorporate modern technologies for data analysis and quality evaluation. The term "Quality 4.0" refers to the ongoing process of digitizing the industry, which makes use of current technology to enhance the quality of both products and services. Industry 4.0 uses Quality 4.0 as a standard to measure itself against, since the digitization of quality management is required to achieve Quality 4.0. The necessity of digitizing high-quality technology, processes, and people as well as the importance of doing so is increasing rapidly. To increase productivity, make quick decisions based on data, including all stakeholders, and provide visibility and accountability, it blends standard quality equipment with collaboration, intelligence, and automation [5].

- As a result of the 4th industrial revolution (Industry 4.0), eleven quality 4.0 axes may be utilized by enterprises to teach, plan, and take action. In addition to this, the framework offers a perspective with regard to traditional consistency. Existing strategies for quality control are not rendered obsolete by Quality 4.0; rather, it enhances and improves upon those strategies as shown in Figure 4.9. The framework should be utilized by manufacturers to conduct an analysis of the current state of their operations and to decide the kinds of enhancements that are required to proceed. These are the following:

- The management of quality relies heavily on this particular aspect of the data. The fourth industrial revolution, or Industry 4.0, enables businesses of any size to have real-time visibility into the effectiveness of their many activities, including manufacturing, production, engineering, and customer service. The speedy and effective collecting of data from a wide variety of sources to facilitate an agile decision-making process is an essential component of Quality 4.0 [6].

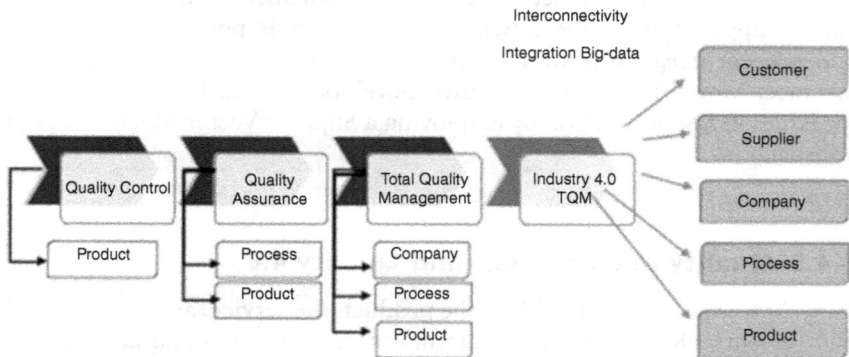

Figure 4.9 Quality management and Quality 4.0.

- Connectivity is a term that refers to the devices and sensors that are employed as part of the Industry 4.0 initiative in manufacturing plants in order to improve live interaction between operational technologies and information communication technologies.
- Collaboration: Both Industry 4.0 and Quality 4.0 are geared toward applying technology to build relationships with customers and examine their levels of satisfaction as well as other feedback they provide [7].
- Compliance: Quality 4.0 offers firms assistance in analyzing the company's existing compliance policies, locating areas in which improvements can be made, and further developing new strategies.
- Culture: Because Industry 4.0 integrates data and processes, it fosters more visibility, interconnection, and collaborative effort. Additionally, it provides the opportunity to make the company culture more plausible and believable [8].
- Development of Apps: This is a valuable and useful tool that helps connect users and industries in order to collect valuable data and feedback to improve the quality of services. Apps can also be thought of as digital business cards. The Fourth Industrial Revolution offers a great deal of promise for the creation and development of new beneficial applications.
- Analytics: Industry 4.0 makes it possible for us to collect enormous amounts of real-time data from the production facility and evaluate it using a variety of technologies.
- Management Systems: One of the more positive aspects of Industry 4.0 is that it allows for the automation of all operations through the use of a variety of software, each of which may be connected to other management systems. This frees up a lot of time for managers to concentrate on innovation by cutting down the amount of time they have to spend on the actual implementation of processes [9].
- With the use of Industry 4.0 technologies, such as cloud computing, software as a service (SaaS), the internet of things (IoT), platform-as-a-service (PaaS), infrastructure as a service (IaaS), etc., scalability can be achieved.
- Industry 4.0 makes leadership more attainable by linking all processes, data, and analytics. This makes it possible to develop the ideal quality culture throughout a company from the top down. Increased visibility, communication, and teamwork are all things that will benefit leaders.
- Proficiency: Industry 4.0 encompasses a variety of elements designed to improve one's level of competency [10]. This includes the usage of social media platforms, which may be utilized to share thoughts and ideas not only across departments but also between corporations. Systems that use AI and ML can make it possible to learn new skills by utilizing virtual reality (VR) and artificial reality (AR) as shown in Figure 4.10.

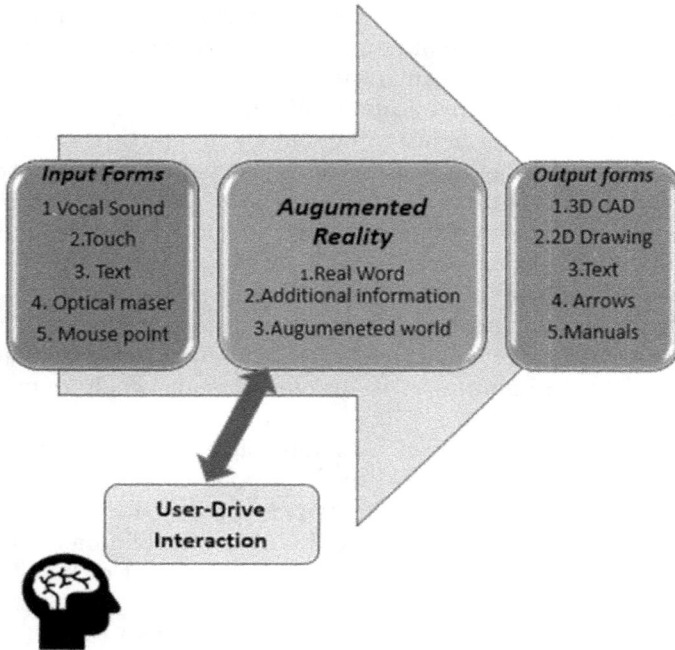

Figure 4.10 Quality Management and Quality 4.0 systems that use AI and ML can make it possible to learn new skills by utilizing virtual reality (VR) and artificial reality (AR).

4.5 PROCESS AUTOMATION AND CONTROL

With the rapid development of information technology, especially network technology, the continuous application of artificial intelligence theory in industrial production and automatic control is boosted [11]. With the assistance of technology that uses AI, any issues that can be identified in their entirety and resolved methodically in the process of industrial automation control. As a result, the process of industrial automation control will become more intelligent, unmanned, and autonomous. The method of performance prediction can make use of some of the most important ML techniques, such as Logistics Regression, Gradient Boosting, and Random Forest. One of the benefits of automating a process is the introduction of predictive maintenance techniques [12]. These techniques allow for the identification and prediction of potential problems in advance of their occurrence, which results in a significant cost reduction in both production and service. The use of ML on sensor data streams is necessary for data-driven predictive maintenance because it enables the identification of data patterns that are symptomatic of future problems and makes it feasible to recognize such patterns [13]. A neural network control platform, an expert decision system and fuzzy control technology, are all examples of AI approaches that are utilized in the building of industrial automatic control systems [14].

4.5.1 Neural network control platform

AI technology simulates the sensing, transmission, and stress processes of the animal nervous system to create a control platform for a neural network. The platform's many interconnected functional parts and networks provide a network topology resembling a neural network as a result of massive parallelization and integration. Data are processed via information gathering, decision-making and processing, based on replicating a biological neural network. Platform processing has improved even more, and the fact that different parts and systems in the network are spread out is now increasingly obvious [15]. In addition, the ability to receive signals and identify information is more accurate, and large-scale storage equipment, cloud platforms, and the new generation of the Internet make it possible to make an interactive and accurate copy of a neural network. The industrial automatic control system can raise the weight coefficient and dynamic level and achieve automatic control of industrial production and processing if it can drive and make decisions on its own without any supervision or help [16].

4.5.2 Fuzzy control technology

It is difficult to build a more accurate mathematical model in the process of industrial automation control given the complicated nature of the process that is being controlled due to the time-varying process. This is because numerous factors influence the process. Therefore, the control technology of replicating human thinking may be carried out with the help of fuzzy logic control, and this can be done without depending on an object model [17]. The fuzzy control system is responsible for gathering, analyzing, and manipulating the data that is collected from the industrial environment. It can send the acquired data, in analog signal form, back to the device that is being controlled. The fundamental component of AI technology is the fuzzy controller. In the context of the application of industrial automation control systems, a variety of controllers are utilized to collect the data of the control object [18]. These controllers are selected following the configuration of the system and the features of the control object. The fuzzy controller serves as the focal point of the typical industrial automation system employing such methods. Other components of the system include the detecting device, controlled object, execution structure and data interface [19]. The motor components make up the majority of the executive structure. To carry out the instructions given by the higher computer, many kinds of motors are utilized, and this choice is made contingent on the specific requirements. A further point to consider is that the role of the detecting fuzzy control system device is comparable to that of the sensing device. The function is to transform and gather the signal coming from the controlled object, and then send it to the top computer through the data link [20].

4.5.3 Expert control system

In the field of industrial automation control, an expert control system has several advantages; among the most significant of these are dependable operation, high adaptability, and reasonable processing and flexible control [21, 22]. AI technology has a significant role to play in the area of industrial automation control since automated real-time monitoring and learning are the main components that enable automatic control. These are among the factors that make this position so crucial. The inference structure is kept as straightforward as is practical to satisfy the real-time requirements of the industrial automation control system due to the restricted knowledge base of the industrial structure [23]. The advantages of AI technology can be fully used by the expert control system. It also makes use of ICT to achieve the acquisition and expression of professional knowledge, database development, and positive logical reasoning, among other things. As a result, it benefits industrial automation control systems with real-time detection, quick reaction, logical reasoning, and self-healing. By combining pertinent models with the workings of AI technology, it is feasible to implement it in the creation of an expert control system based on AI [24]. Depending on how easily the parameters are adjusted, complex system logic reasoning may also be realized. In addition, classical logical reasoning is the foundation upon which the current level of AI technology is built [24]. The precision and rationalism of the system can be maximized using this supposition, which may also be maximized in industrial automation control systems using the artificial intelligence technology

4.6 CONCLUSION

This chapter provides an overview of how AI can be used to automate industrial data performance. By using machine learning to ensure the optimization of their production processes, manufacturing plants improve both the efficiency of their operations and the quality of their products. This assists manufacturers in keeping their advantage over their competitors. It is a cutting-edge technology since it can be used to predict how well production systems will operate thanks to the ML application. It is possible to significantly increase the output of industrial processes by first modeling the various possible operating situations and then modeling the control parameters. In addition, it is now feasible to reduce expenditures in terms of both time and money while maintaining the same level of success. The most recent advancements in the field of Industry 4.0 have made it possible for AI and ML to be applied in production as well as research. This is an exciting new development. The application of AI and ML in industrial settings is dissected in detail in this chapter. The discussion has taken place regarding the use of AI and ML in decision-intensive activities such as PPC. The uses of ML and AI and in PPC tasks can be broken down into three key areas: performance evaluation, dynamic scheduling and monitoring, and automated process control. Industry 4.0 has made tremendous progress feasible in these three application sectors, demonstrating the importance of this technological advancement. This has made it feasible to improve the automation of human-dependent processes that commonly need complicated decision-making on the part of the production and process supervisors. This has made it possible to improve the automation of human-dependent processes. As a result of the automation process, there is no longer any room for error on the part of humans. Performance evaluation and monitoring make it possible to carry out activities such as the dynamic real-time assignment of resources and tasks, real-time production planning, and analytics. This can boost the productivity of the industrial process as a whole. Last but not least, the automation of processes will make it feasible to automate PPC. This opens the door to considerable improvements in the decision-making process during production planning, which, in turn, makes production planning much more effective. It is possible to conclude that ML plays a significant part in performance prediction, scheduling, and automation and that it will also continue to be employed in other main application fields. In the day and age of Industry 4.0, it is beyond any doubt that manufacturing facilities are going to receive a great number of additional benefits that they did not anticipate from machine learning and artificial intelligence.

REFERENCES

[1] MacCarthy, B. L., & Liu, J. (1993). Addressing the gap in scheduling research: A review of optimization and heuristic methods in production scheduling. *International Journal of Production Research*, 31(1), 59–79.

[2] Cowling, P. I., & Johansson, M. (2002). Using real-time information for effective dynamic scheduling. *European Journal of Operational Research*, 139(2), 230–244.

[3] Bustillo, A., Grzenda, M., & Macukow, B. (2016). Interpreting tree-based prediction models and their data in machining processes. *Integrated Computer-Aided Engineering*, 23(4), 349–367.

[4] Li, W., Li, H., Gu, S., & Chen, T. (2020). Process fault diagnosis with model- and knowledge-based approaches: Advances and opportunities. *Control Engineering Practice*, 105, 104637.

[5] Büchi, G., Cugno, M., & Castagnoli, R. (2020). Smart factory performance and Industry 4.0. *Technological Forecasting and Social Change*, 150, 119790.

[6] Kumar, S.S., Sudhir Bale, A., & Matapati Vinay, P.M. (2021). Conceptual study of artificial intelligence in smart cities with industry 4.0. In *Proceedings of the 2021 International Conference on Advance Computing and Innovative Technologies in Engineering (ICACITE)*, Greater Noida, India, 4–5 March 2021.

[7] Oluyisola, O.E., Bhalla, S., Sgarbossa, F., & Strandhagen, J.O. (2022). Designing and developing smart production planning and control systems in the industry 4.0 era: A methodology and case study. *Journal of Intelligent Manufacturing*, 33, 311–332.

[8] Kamble, S.S., Gunasekaran, A., Ghadge, A., & Raut, R. (2020). A performance measurement system for industry 4.0 enabled smart manufacturing system in SMMEs- A review and empirical investigation. *International Journal of Production Economics*, 229, 107853.

[9] Kazi, M.-K., Eljack, F., & Mahdi, E. (2021). Data-driven modeling to predict the load vs. displacement curves of targeted composite materials for industry 4.0 and smart manufacturing. *Composite Structures*, 258, 113207.

[10] Oztemel, E., & Gursev, S. (2020). Literature review of Industry 4. 0 and related technologies. *Journal of Intelligent Manufacturing*, 31, 127–182.

[11] Ammar, M., Haleem, A., Javaid, M., Walia, R., & Bahl, S. (2021). Improving material quality management and manufacturing organizations system through Industry 4. 0 technologies. *Materials Today*, 45, 5089–5096.

[12] Sahal, R., Alsamhi, S., Breslin, J., Brown, K., & Ali, M. (2021). Digital twins collaboration for automatic erratic operational data detection in Industry 4.0. *Applied Sciences*, 11, 3186.

[13] Braz, R.G.F., Scavarda, L.F., & Martins, R.A. (2011). Reviewing and improving performance measurement systems: An action research. *International Journal of Production Economics*, 133, 751–760.

[14] Atik, H., & Ünlü, F. (2019). The measurement of Industry 4.0 performance through Industry 4.0 index: An empirical investigation for Turkey and European Countries. *Procedia Computer Science*, 158, 852–860.

[15] Nenadál, J. (2020). The new EFQM model: What is new and could be considered as a suitable tool concerning quality 4.0 concept? *Quality Innovation Prosperity*, 24, 17.

[16] Frederico, G.F., Garza-Reyes, J.A., Kumar, A., & Kumar, V. (2020). Performance measurement for supply chains in the Industry 4.0 era: A balanced scorecard approach. *International Journal of Production Performance Management*, 70, 789–807.

[17] Kaggle. (2016). "Bosch Production Line Performance." Available: https://www.kaggle.com/c/bosch-production-line-performance

[18] Davis, J., Edgar, T., Graybill, R., et al. (2015). Smart manufacturing. *Annual Review of Chemical and Biomolecular Engineering*, 6, 141–160.

[19] Adesiyan, Ayomide. (2021). Performance prediction of production lines using machine learning algorithm. *ScienceOpen Preprints*. DOI: 10.14293/S2199-1006.1.SOR-.PPA7BE8.v1

[20] Wu, D., & Mendel, J. M. (2019). Recommendations on designing practical interval type-2 fuzzy systems. *Engineering Applications of Artificial Intelligence*, 85, 182–193.

[21] Mendel, J. M. (2018). *Uncertain Rule-Based Fuzzy Logic Systems: Introduction and New Directions*, 2nd edition, Springer International Publishing AG, Cham, Switzerland.

[22] Lu, Z. Y., Wang, M. Q., Dai, W., & Sun, J. H. (2019). In-process complex machining condition monitoring based on deep forest and process information fusion. *The International Journal of Advanced Manufacturing Technology*, 104(5–8), 1953–1966.

[23] Xia, M., Li, T., Shu, T. et al. (2019). A two-stage approach for the remaining useful life prediction of bearings using deep neural networks. *IEEE Transactions on Industrial Informatics*, 15(6), 3703–3711.

[24] Tieng, H., Tsai, T.-H., Chen, C.-F., Yang, H.-C., Huang, J.-W., & Cheng, F.-T. (2018). Automatic virtual metrology and deformation fusion scheme for engine-case manufacturing. *IEEE Robotics and Automation Letters*, 3(2), 934–941.

Chapter 5

Data-driven optimization on the workability and strength properties of M-30 grade concrete using MOORA

A. Anandraj, S. Vijayabaskaran and P.V. Rajesh

Saranathan College of Engineering, Panjappur, Tamilnadu, India

CONTENTS

5.1 INTRODUCTION

A significant number of studies have been done in search of suitable, accessible and cheaper-to-produce materials. Blended cements along with the use of recycled material as aggregate substitute are rising rapidly in the construction industry. Since the construction industry necessitates a steady supply of construction materials, which must be derived from natural virgin resources, there will be a scarcity of construction materials and the cost of materials and

DOI: 10.1201/9781003257714-5

construction will rise (Adaway and Wang, 2015). In the event of a shortage, researchers have added a cheap and readily available substitute such as bentonite in order to receive more attention from customers and greater use of industrial end products in concrete as aggregates. The incorporation of these materials is of great benefit, not only with regard to the environmental and energy efficiency elements of the cement industry, but also with regard to the structural durability, reliability, integrity and lifecycle cost aspects.

Mostly made up of smectite clay, volcanic ash clay, known as bentonite, forms when it is altered. This item contains additional minerals such as quartz, calcite, feldspar, and gypsum, which may also include montmorillonite. When water mixes with bentonite, its volume increases about sixfold, creating a viscous, gelatinous liquid. The varied uses of bentonite, which include water absorption, swelling, viscosity, and thixotropy, make it a valuable material for a variety of uses and applications in construction industry (Afzal et al., 2014).

In ordinary Portland cement, bentonite helps to promote pozzolanic action, making the cement more fluid. A pozzolano is a siliceous and aluminous material that reacts with calcium hydroxide at a normal temperature in order to form cementitious compounds. When water is added to a mixture of OPC and pozzolano, the silica in the water interacts with the calcium hydroxide in hydrated cement paste, resulting in an increased amount of soluble silica in the paste. The reaction does not produce lime, but rather consumes it, making the hydrated cement paste in acidic environments even more durable (Corinaldesi et al., 2005). Hydrated pozzolanic cements yield highly reactive reaction products that efficiently fill larger capillary pores. This boosts concrete's impermeability and strength.

Glass has been one of the earliest-known materials that humans have ever made. Recycling these forms instead of stockpiling or sending them to landfills reduces the levels of environmental damage. Practically, glass can be either recycled or reused completely. There is no loss of quality as the process is repeated indefinitely. There are many ways of successful recycling of waste glass into cullet glass, such as for sand blasting, for abrasives, in concrete, and for pozzolanic additives in concrete. The recycling and reusing of industrial by-products and wastes, such as waste glass, has produced gains in the construction sector (Shekhawat and Aggarwal, 2014). Recycling of this waste, which is usually destined for landfill, can instead be particularly used to create fine aggregate. Doing so results in a reduction in the use of landfill and also the demand for natural raw material extraction for construction activities.

In terms of both ecology and economics, using Recycled Concrete Aggregate instead of normal aggregates has a positive impact. It can reduce the need to open new quarries, preserving the environment and reducing the enormous consumption of fuel/energy associated with the hauling of mined materials. Over similar hauling distances, the energy required for the transportation of recycled concrete aggregate is lower than that of normal concrete aggregate, since the unit weight of recycled concrete aggregate is lighter than that of normal aggregate (Verian et al., 2018). At the same time, Recycle Concrete Aggregate (RCA) reduces the generation of construction wastes and helps in their removal. Construction costs may also be reduced through the use of RCA. Research conducted by the Environmental Council of Concrete Organizations reveals that replacing normal aggregate with RCA can save up to 60% (Abbas et al., 2009). It has been discovered that using coarse Recycle Concrete Aggregate derived from building demolition waste minimizes green-house gas emissions by up to 65%, while also saving up to 58% on energy consumption (Afroughsabet et al., 2017). Other similar studies have also sug-gested that concrete made up of RCA can be formulated in such a way as to match the quality of concrete produced with normal aggregate, thereby avoid-ing the use of extra cement or any other additive materials.

5.2 MATERIALS SPECIFICATIONS AND METHODOLOGY

5.2.1 Material specifications

Several materials, including Cement, Fine aggregate, Coarse aggregate, Bentonite, Recycled Glass Aggregate, Recycled Concrete Aggregate, Water, and Super-Plasticizer, were used in the research study. Throughout this work, Ordinary Portland Cement (OPC) of grade 43 was used. Based on the basic test results, Specific gravity was 3.17, Initial setting time is given as 42 minutes and Final setting time is given as 575 minutes. Clean river sand was allowed to pass through a 4.75 mm IS sieve before being used in this investigation. Fineness modulus and Specific gravity were both found to be 2.65 and 2.74. Aggregate obtained from local quarry is used in concrete production as Coarse aggregate. (Coarse aggregate had a maximum nomi-nal size of 20mm.) Specific gravity, fineness modulus and water absorption of coarse aggregate was ascertained to be 2.76, 5.2 and 0.92% respectively. Yellowish sodium bentonite powder, with specific gravity 2.35, is used. Local glass market's waste glass is used as Recycled Glass Aggregate which had a specific gravity value of 2.18 and a fineness modulus index of 2.36. The RCA, which is used here, had been obtained from a nearby site, where a 12-year-old building had been demolished. The aggregate exhibited a spe-cific gravity, fineness modulus and water absorption of Coarse aggregate was found to be 2.67, 2.72 and 1.15% respectively.

5.2.2 Methodology

The experimental programme involved an attempt to research concrete's strength properties with bentonite, recycled glass aggregate and recycled coarse aggregate in place of cement and natural aggregates. In this research 16 concrete mix samples, with various percentages of bentonite, recycled

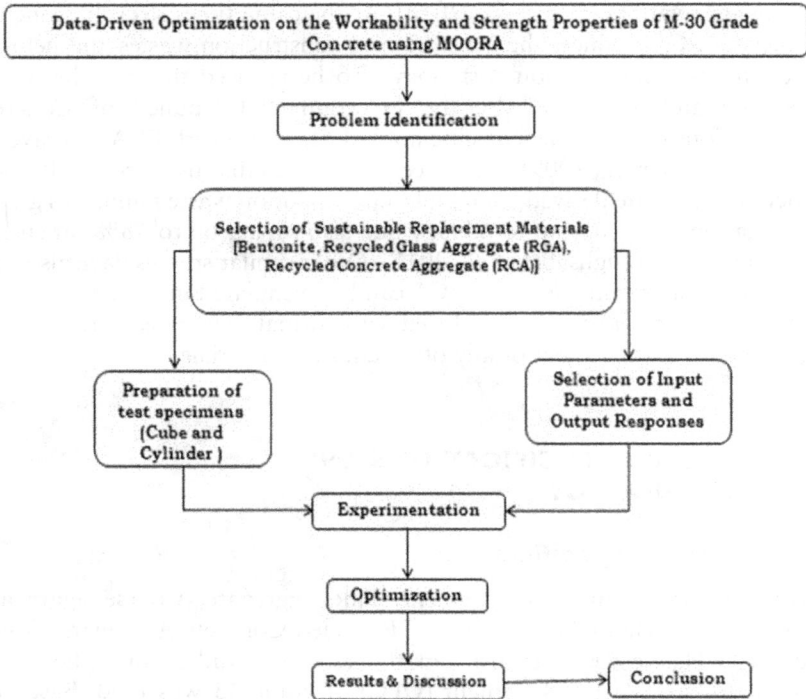

Figure 5.1 Schematic representation of research methodology.

glass aggregate and recycled coarse aggregate, have been used extensively to evaluate the characteristics of the resulting concrete. The concrete mixes used to investigate different percentages of concrete using bentonite, RGA and RCA were labelled consecutively as TM 01, TM 02,...,TM 16, where TM stands for Trial Mix. Figure 5.1 indicates the procedural methodology for conducting the experimental research.

5.3 EXPERIMENTAL PROCEDURE

5.3.1 Mix proportioning

Using IS-10262:2019 as a guide, a mix design was developed. This standard provides the proportioning and preparation guidelines for concrete mixes, depending on the concrete-making materials, in order to comply with specifications. Two types of mixes were prepared; one used cement, fine aggregate, and coarse aggregate in a 1:1:67:2.84 ratio, and the other replaced cement with bentonite (0%, 5%, 10%, 15%), fine aggregate with recycled glass (0%, 10%, 15%, 20%), and coarse aggregate with recycled concrete (0%, 10%, 15%, 20%). Both mixes had a water-to-cement ratio of 0.45.

5.3.2 Preparation of specimens

For the compressive strength, a cube mould $150 \times 150 \times 150$ mm was used. To acquire split tensile strength, cylindrical mould with diameter 150mm and length 300mm was employed. Because the moulds were made of iron, it is easy to extract the finished specimen. Both moulds comes with a metal base plate with a flat surface. A mineral oil brush was used to keep water from escaping and concrete from bonding with inner walls. The concrete was casted in three layers, each one being 50mm thick. Once the surface of the concrete mould was completely smooth, the concrete was vibrated for between one and one-and-a-half minutes to remove cavitation bubbles from the surface. Moulds were left to dry for 24 hours. Once they were dry, the concrete specimens were pulled out of the moulds and immersed in fresh water, where it was cured for testing.

5.4 TESTS, RESULTS AND DISCUSSIONS

5.4.1 Slump test

The "slump" or "workability" of concrete refers to the fluidity of fresh concrete before it sets. It is assumed that higher the slump, more fluid the concrete. It's a somewhat perplexing phrase for a seemingly complex procedure, but when it comes down to the result, it's actually an easy reference factor. The concrete slump test is a method of determining the consistency of various concrete mixes. The slump test is used to determine the concrete's workability and flowability, and also to detect a bad mix.

Slump of all mixes with constant water-cement (w/c) ratio for the same group was measured in this experiment to learn about workability changes caused by bentonite, recycled glass aggregate, and recycled concrete aggregate. Table 5.1 shows that when concrete is mixed with cement substituted bentonite and aggregates made from recycled materials, the slump is reduced when compared to the control mix. The lower the slump, the higher the workability (slump and workability are inversely proportional).

A graph has been plotted taking specimens with different mix IDs on the x-axis and slump values on the y-axis. It clearly shows in Figure 5.2 that the workability of the mixes is following a gradual decrease.

5.4.2 Compressive strength

In concrete, the compressive strength is defined as the strength of hardened concrete as measured by the compression test method. It is possible to determine the compression strength of concrete by measuring its ability to bear compressive stresses. This is calculated by compressing cubic concrete specimens in a compression testing equipment to obtain the required strength.

Table 5.1 Workability test results of all mixes

MixID	Partial replacement of cement by bentonite (%)	Partial replacement of fine aggregate by recycled glass aggregate (%)	Partial replacement of coarse aggregate by recycled concrete aggregate (%)	Slump(mm)
TM 01	0	0	0	135
TM 02	0	10	10	132
TM 03	0	15	15	124
TM 04	0	20	20	121
TM 05	5	0	0	120
TM 06	5	10	10	118
TM 08	5	20	20	108
TM 09	10	0	0	110
TM 10	10	10	10	107
TM 11	10	15	15	105
TM 12	10	20	20	103
TM 13	15	0	0	96
TM 14	15	10	10	88
TM 15	15	15	15	78
TM 16	15	20	20	70

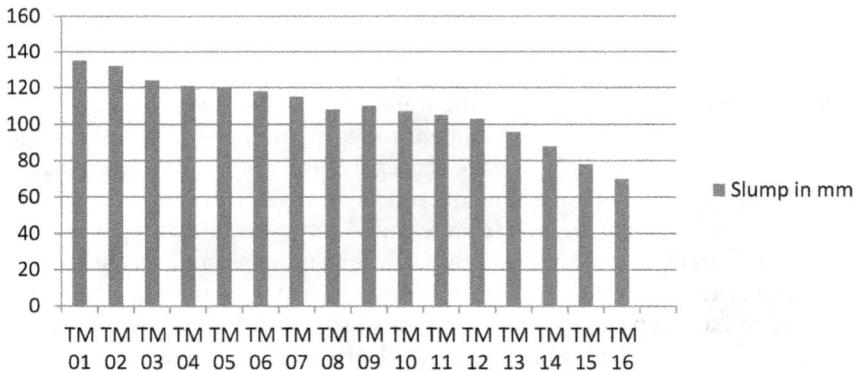

Figure 5.2 Comparison of workability among different mixes.

This method is used to assess the concrete's strength and integrity. If this work was done correctly, this one test will reveal whether or not the job was completed appropriately. When it comes to compressive strength, a number of factors come into play, including the water-to-cement ratio, cement strength, concrete material quality, and quality control during the concrete-fabrication process.

Figure 5.3 Testing for compressive strength.

The compressive strength test is done by placing the cube centrally between the flat pads at the top and the bottom of the Compression Testing Machine and applying load, as shown in Figure 5.3. The test is performed on a 150mm × 150mm × 150mm cube. When the specimen is crushed, the crushing load is recorded, and the crushing load divided by specimen crushing area yields concrete compressive strength (Crushing Load/Crushing Area). In this case, compressive strength is the average of the strengths of three cubes measured at various points throughout the curing process.

5.4.3 Tensile strength split test

The Tensile Strength Split Test was completed by applying the load to a horizontally placed cylinder between the flat pads of the Compression Testing Machine (Figure 5.4). To ensure uniform load distribution, plywood strips are provided above and below the specimen where it comes into contact with the plate. The load is applied continuously without shock, and the cylinder's ultimate splitting load is recorded.

Table 5.2 shows the average compressive and split tensile strengths (C_S and T_S) of casted cubes and cylinders determined according to IS 516-1959 using bentonite, recycled glass aggregate, and recycled concrete aggregate at 7 and 28 days. In the table, C stands for Cement, Be stands for Bentonite, FA stands for fine aggregate and CA stands for coarse aggregate.

A graph has been plotted taking specimens with different mix IDs on the x-axis and compressive strength values after 7 and 28 days on the y-axis. It has been clearly illustrated in Figure 5.5 that the compressive strength test values follow a random distribution pattern with respect to the curing days and various mix ratios.

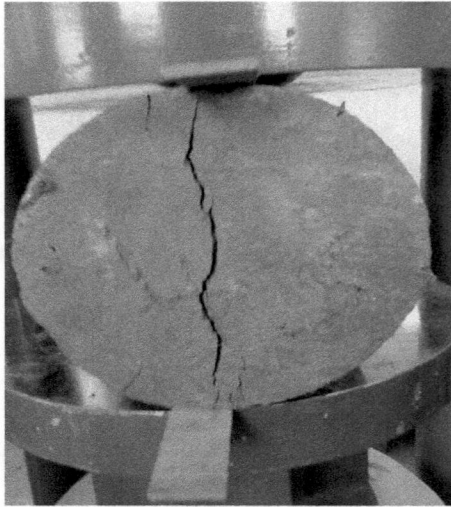

Figure 5.4 Testing for split tensile strength.

Table 5.2 Strength test results of all mixes

Mix TM ID	C (%)	Be (%)	FA (%)	RGA (%)	CA (%)	RCA (%)	C_s (N/mm²)		T_s (N/mm²)	
							7Days	28Days	7Days	28Days
01	100	0	100	0	100	0	25.46	41.18	3.82	6.18
02	100	0	90	10	90	10	25.54	40.64	3.83	6.10
03	100	0	85	15	85	15	27.36	41.85	4.10	6.28
04	100	0	80	20	80	20	24.28	35.46	3.64	5.32
05	95	5	100	0	100	0	15.58	36.68	2.34	5.50
06	95	5	90	10	90	10	14.75	35.72	2.21	5.36
07	95	5	85	15	85	15	14.67	34.67	2.20	5.20
08	95	5	80	20	80	20	15.09	35.86	2.26	5.38
09	90	10	100	0	100	0	18.46	41.45	2.77	6.22
10	90	10	90	10	90	10	20.48	41.12	3.07	6.17
11	90	10	85	15	85	15	16.22	39.35	2.43	5.90
12	90	10	80	20	80	20	14.45	37.42	2.17	5.61
13	85	15	100	0	100	0	15.78	38.67	2.37	5.80
14	85	15	90	10	90	10	13.68	36.28	2.05	5.44
15	85	15	85	15	85	15	13.33	33.45	2.00	5.02
16	85	15	80	20	80	20	12.34	28.46	1.85	4.27

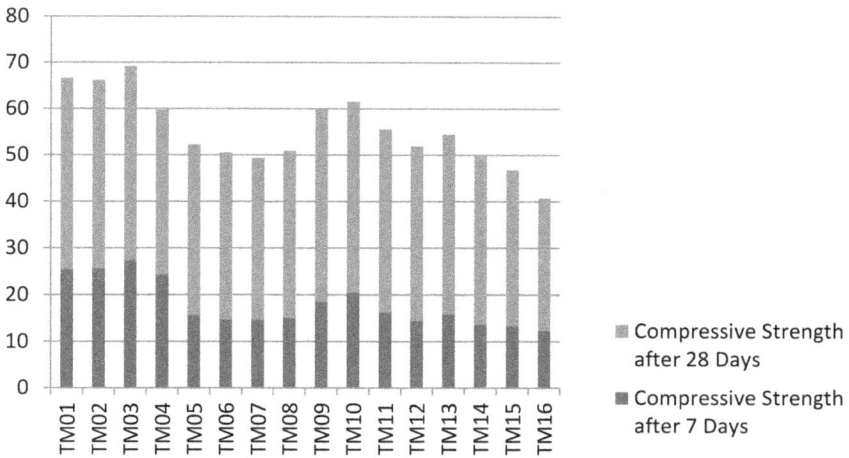

Figure 5.5 Comparison of C_s after 7 days and 28 days.

Another graph, shown as Figure 5.6, has been plotted taking specimens with different mix ID's on the x-axis and split tensile strength values after 7 and 28 days on the y-axis. It has been clearly demonstrated that the split tensile strength test values also follow a random distribution pattern with respect to the curing days and various mix ratios.

Since all the parameters, including workability, compressive strength after 7 days as well as 28 days and split tensile strength after 7 days as well as 28 days, are following a randomly distributed pattern it is necessary to find an ideal mix ratio has to be found out. To help in this it is found that

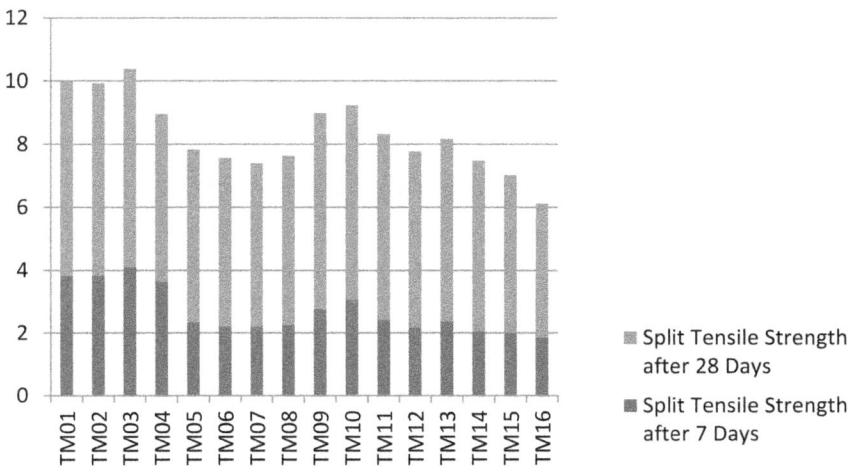

Figure 5.6 Comparison of T_S after 7 days and 28 days.

Table 5.3 Constraints for optimization

Type	Name of the variables	Goal	Lower limit	Upper limit
Factors/ Parameters	Proportion of Cement	Within range	85	100
	Proportion of Fine Aggregate	Within range	80	100
	Proportion of Coarse Aggregate	Within range	80	100
Properties/ Responses	Slump	Minimize	70	135
	Compressive Strength after 1 week	Maximize	12.34	27.36
	Compressive Strength after 1 month	Maximize	28.46	41.85
	Tensile Strength after 1 week	Maximize	1.85	4.1
	Tensile Strength after 1 month	Maximize	4.27	6.28

the effective way is going for optimization techniques. In Table 5.3, various constraints for optimization has been formulated.

5.5 OPTIMIZATION

The inseparable relationship between optimization and research is a requirement nowadays. In the method of optimization, you seek the optimal result, which is most probably the best feasible given the set of options available while considering all of the parameters of the process or testing set. In the world of optimization, it is possible to pursue one of two goals: to maximize or to minimize the optimal solution (Kou et al., 2012). In its most basic form, an optimization problem will be maximizing or minimizing a real function.

5.5.1 Multi-Criteria Decision-Making (MCDM)

Multi-Criteria Decision-Making (MCDM) is a sub-branch of optimization in which multiple criteria are used to determine the optimal solution from a combinational set of alternatives/choices. It is gaining its importance in terms of resource, transportation, material inventory, production planning control and management. The real-life decision-making problems that we encounter are always comprised of multiple conflicting criteria and objectives, all of which must be taken into consideration simultaneously. For example, the compromises needed to balance the performance and quality study hours of a student or the performance and quality of teaching methodology of a teacher. The choice of materials is subject to similar conflicts between material properties and performance metrics (Brauers et al., 2008). The choice of materials is certainly one of the most important of the many fields where MCDM is applied. The selection of a new material or the replacement of an existing one with a different material, with better performance characteristics, is usually done using trial-and-error methods or by means of previous experience. This may or may not lead to an optimal design solution, but the use of the MCDM method will contribute to preventing the use of inappropriate materials and to minimizing cost factors.

5.5.2 Multi-objective optimization based on ratio analysis (MOORA)

Investigators of complex problems to discover fruitful solutions for selection-based problems are finding it increasingly difficult to take crucial decisions in the real-time construction environment. Each alternative's objectives and outcomes must be both tangible and measurable. Some of the competing criteria are beneficiary in nature (a maximized solution is required) and some are detrimental (a minimized solution is required). The multi-objective optimization based on ratio analysis (MOORA) method takes into account both beneficiary and non-beneficiary objectives, where ranking is done to identify the optimal alternative from a set of available substitutes. This technique is better suited for problems that have the attributes of both maximizing and minimizing (Karande and Chakraborty, 2012).

5.5.3 Procedural steps in MOORA

This method initiates with a Decision Matrix (DM) indicated in Equation (5.1), which demonstrates the performance of various alternatives in relation to a variety of criteria.

$$
\begin{bmatrix}
x_{11} & x_{12} & \cdots & x_{1n} \\
x_{21} & x_{22} & \cdots & x_{2n} \\
\vdots & \vdots & \vdots & \vdots \\
x_{m1} & x_{m2} & \cdots & x_{mn}
\end{bmatrix}
\tag{5.1}
$$

where x_{ij} denotes the confluence of the ith alternative with j^{th} criterion, m denotes the number of choices presented row-wise, and n denotes the number of criteria represented column-wise. The Decision Matrix, thus formed is indicated in Table 5.4.

The decision matrix is then normalised, so that it becomes a dimensionless quantity. The following simple normalisation procedure has been

Table 5.4 Decision matrix

	Process variables/Factors			Criterial/Properties				
Mix ID	A:Cement	B:Fine aggregate	C:Coarse aggregate	Slump	Compressive strength after 7 days	Compressive strength after 28 days	Tensile strength after 7 days	Tensile strength after 28 days
	%	%	%	Mm	N/mm²	N/mm²	N/mm²	N/mm²
1	100	100	100	135	25.46	41.18	3.82	6.18
2	100	90	90	132	25.54	40.64	3.83	6.1
3	100	85	85	124	27.36	41.85	4.1	6.28
4	100	80	80	121	24.28	35.46	3.64	5.32
5	95	100	100	120	15.58	36.68	2.34	5.5
6	95	90	90	118	14.75	35.72	2.21	5.36
7	95	85	85	115	14.67	34.67	2.2	5.2
8	95	80	80	108	15.09	35.86	2.26	5.38
9	90	100	100	110	18.46	41.45	2.77	6.22
10	90	90	90	107	20.48	41.12	3.07	6.17
11	90	85	85	105	16.22	39.35	2.43	5.9
12	90	80	80	103	14.45	37.42	2.17	5.61
13	85	100	100	96	15.78	38.67	2.37	5.8
14	85	90	90	88	13.68	36.28	2.05	5.44
15	85	85	85	78	13.33	33.45	2	5.02
16	85	80	80	70	12.34	28.46	1.85	4.27

followed in this process. It is significantly noted that the decision matrix's elements are normalised without taking the type of criteria into account (beneficial or non-beneficial). The normalised value for any criterion should not exceed one, as indicated in Equations (5.2) and (5.3) respectively. It should be between −1 and +1.

$$r_{ij} = \left[2^* \left\{ \frac{x_{ij} - \min(x_i)}{\max(x_i) - \min(x_i)} \right\} \right] - 1 \text{ (for maximizing criteria elements)} \quad (5.2)$$

$$r_{ij} = \left[2^* \left\{ \frac{\mathrm{man}(x_i) - x_{ij}}{\max(x_i) - \min(x_i)} \right\} \right] - 1 \text{ (for manimizing criteria elements)} \quad (5.3)$$

where,
   i values ranging from 1, 2..., m and j values ranging from 1, 2..., n.
   The normalization matrix R is denoted in Equation (5.4) and the respective values were given in Table 5.5.

Table 5.5 Normalized Decision Matrix

Slump	C_s after 7 days	C_s after 28 days	T_s after 7 days	T_s after 28 days
mm	N/mm²	N/mm²	N/mm²	N/mm²
−1.000	0.747	0.900	0.751	0.900
−0.908	0.758	0.819	0.760	0.821
−0.662	1.000	1.000	1.000	1.000
−0.569	0.590	0.046	0.591	0.045
−0.538	−0.569	0.228	−0.564	0.224
−0.477	−0.679	0.084	−0.680	0.085
−0.385	−0.690	−0.072	−0.689	−0.075
−0.169	−0.634	0.105	−0.636	0.104
−0.231	−0.185	0.940	−0.182	0.940
−0.138	0.084	0.891	0.084	0.891
−0.077	−0.483	0.627	−0.484	0.622
−0.015	−0.719	0.338	−0.716	0.333
0.200	−0.542	0.525	−0.538	0.522
0.446	−0.822	0.168	−0.822	0.164
0.754	−0.868	−0.255	−0.867	−0.254
1.000	−1.000	−1.000	−1.000	−1.000

$$\begin{bmatrix} r_{11} & r_{12} & \cdots & r_{1n} \\ r_{21} & r_{22} & \cdots & r_{2n} \\ \vdots & \vdots & \vdots & \vdots \\ r_{m1} & r_{m2} & \cdots & r_{mn} \end{bmatrix} \tag{5.4}$$

Each criterion should be having a dedicated weight factor. In this research work, all of the criteria are given equal weightage of 20% (weight factor = 0.2).

All criteria values should be multiplied by the weightage, as represented in Equation (5.5):

$$a_{ij} = w_j * r_{ij} \tag{5.5}$$

The resultant of the weighted normalized DM is given in Equations (5.6) and Table 5.6:

Table 5.6 Weight-normalized decision matrix

Slump	C_s after 7 days	C_s after28 Days	T_s after7 Days	T_s after28 Days
mm	N/mm²	N/mm²	N/mm²	N/mm²
−0.200	0.149	0.180	0.150	0.180
−0.182	0.152	0.164	0.152	0.164
−0.132	0.200	0.200	0.200	0.200
−0.114	0.118	0.009	0.118	0.009
−0.108	−0.114	0.046	−0.113	0.045
−0.095	−0.136	0.017	−0.136	0.017
−0.077	−0.138	−0.014	−0.138	−0.015
−0.034	−0.127	0.021	−0.127	0.021
−0.046	−0.037	0.188	−0.036	0.188
−0.028	0.017	0.178	0.017	0.178
−0.015	−0.097	0.125	−0.097	0.124
−0.003	−0.144	0.068	−0.143	0.067
0.040	−0.108	0.105	−0.108	0.104
0.089	−0.164	0.034	−0.164	0.033
0.151	−0.174	−0.051	−0.173	−0.051
0.200	−0.200	−0.200	−0.200	−0.200

Table 5.7 Assessment values

Mix TM ID	C (%)	Be (%)	F.A. (%)	RGA (%)	C.A. (%)	RCA (%)	Assessment Value, y_i
01	100	0	100	0	100	0	0.860
02	100	0	90	10	90	10	0.813
03	100	0	85	15	85	15	0.932
04	100	0	80	20	80	20	0.368
05	95	5	100	0	100	0	−0.029
06	95	5	90	10	90	10	−0.143
07	95	5	85	15	85	15	−0.228
08	95	5	80	20	80	20	−0.178
09	90	10	100	0	100	0	0.349
10	90	10	90	10	90	10	0.418
11	90	10	85	15	85	15	0.072
12	90	10	80	20	80	20	−0.150
13	85	15	100	0	100	0	−0.046
14	85	15	90	10	90	10	−0.352
15	85	15	85	15	85	15	−0.599
16	85	15	80	20	80	20	−1.000

$$\begin{bmatrix} a_{11} & a_{12} & \cdots & a_{1n} \\ a_{21} & a_{22} & \cdots & a_{2n} \\ \vdots & \vdots & \vdots & \vdots \\ a_{m1} & a_{m2} & \cdots & a_{mn} \end{bmatrix} \tag{5.6}$$

The assessment value is the negation of the total of variables in the non-beneficiary criteria of each row, y_i from the sum of variables in the beneficiary criteria of each row, which is a row-wise calculation (of each row). The assessment value of each alternative is given in Table 5.7.

The best alternative is the one with the highest assessment value when categorized in descending order. To determine the final preference of the specimen alternatives, an order-based ranking of y_i values is recommended. Table 5.8 indicates the ranking of alternatives in the ascending order from best to worst, based on the obtained assessment value.

The optimized solution is represented in Table 5.9.

This method is quite novel and flexible in terms of the methodology involved. This method is also more credible than other existing MCDM methods because it is attribute-based and relies on decisive and perpetual data.

Table 5.8 Ranking of alternatives

Assessment value, y_i	Rank	Alternatives
0.860	2	1
0.813	3	2
0.932*	1*	3*
0.368	5	4
−0.029	8	5
−0.143	10	6
−0.228	13	7
−0.178	12	8
0.349	6	9
0.418	4	10
0.072	7	11
−0.150	11	12
−0.046	9	13
−0.352	14	14
−0.599	15	15
−1.000	16	16

*To highlight the assessment value and alternatives which "Ranked 1".

Table 5.9 Optimized solution

Mix ID No./ Specimen No.	Cement	Fine aggre-gate	Coarse aggre-gate	Slump	Compressive strength after 7 days	Compressive strength after 28 days	Tensile strength after 7 days	Tensile strength after 28 days
	%	%	%	Mm	N/mm^2	N/mm^2	N/mm^2	N/mm^2
1	100	85	85	124	27.36	41.85	4.1	6.28

5.6 CONCLUSION

The main findings from this research are as follows: Replacing cement with bentonite, fine particles with recycled glass, and coarse aggregates with recycled concrete reduces the workability of the concrete mix. Despite this decline in the slump of the aforementioned mixtures, the concrete mix possessed a good workability. The optimum mix was chosen based on MCDM-MOORA. This involves the partial replacement of cement with 0% bentonite, fine aggregate with 15% recycled waste glass, and coarse aggregate with 15% recycled concrete.

The partial replacement cement with 0% bentonite, fine aggregate with 15% recycled waste glass, and coarse aggregate with 15% recycled results in a 6.94% and 1.61% increase in compressive strength after 7 and 28 days of curing respectively when compared to control concrete. This shows that the initial achievement of compressive strength is faster and slowly it decreases when the curing has been prolonged. The fractional replacement of cement with 0% bentonite, fine aggregate with 15% recycled waste glass, and coarse aggregate with 15% recycled concrete results in a 6.83% and 1.58% increase in split tensile strength after 7 and 28 days of curing, respectively, when compared to control concrete. This shows that the initial achievement of compressive strength is at a faster pace and slowly it decreases when the curing has been prolonged. The optimal enhancement in compressive strength was found when partial replacement of cement with bentonite at 10%, partial replacement of natural fine aggregate with recycled glass aggregate at 10% and partial replacement of natural coarse aggregate with recycled aggregate at 10% was 20.48 N/mm²for 7 days and 41.12 N/mm² for 28 days.

The optimal enhancement in split tensile strength was found when the partial replacement of cement with bentonite at 10%, the partial replacement of natural fine aggregate with recycled glass aggregate at 10% and the partial replacement of natural coarse aggregate with recycled aggregate at 10% was 3.07 N/mm²for 7 days and 6.17 N/mm² for 28 days. The use of recycled glass and recycled concrete as a replacement for fine and coarse aggregate resulted in no discernible change in the colour of the concrete. In this effort, it has been figured out how to utilize discarded glass as fine aggregate in concrete and recycled concrete as coarse aggregate in concrete in an efficient manner. The statistics provided in this chapter demonstrate that there is a significant untapped potential for the usage of recycled glass and recycled aggregate in concrete; future research can explore the long-term effect on the characteristics of concrete as a result of the data published in this chapter.

ACKNOWLEDGEMENT

The authors owe a special thanks to the Department of Civil Engineering, Saranathan College of Engineering, Tiruchirappalli, Tamilnadu, INDIA for providing the laboratory facilities to accomplish this research work.

FUNDING

This study was not funded by any funding sources.

CONFLICTS OF INTEREST

The authors declare no conflicts of interest.

BIBLIOGRAPHY

Abbas, A., Fathifazl, G., Isgor, O. B., Razaqpur, A. G., Fournier, B., & Foo, S. (2009). Durability of recycled aggregate concrete designed with equivalent mortar volume method. *Cement and Concrete Composites, 31*(8), 555–563. https://doi.org/10.1016/j.cemconcomp.2009.02.012

Adaway, M., & Wang, Y. (2015). Recycled glass as a partial replacement for fine aggregate in structural concrete—Effects on compressive strength. *Electronic Journal of Structural Engineering, 14*(1), 116–122.

Afroughsabet, V., Biolzi, L., & Ozbakkaloglu, T. (2017). Influence of double hooked-end steel fibers and slag on mechanical and durability properties of high performance recycled aggregate concrete. *Composite Structures, 181*, 273–284. https://doi.org/10.1016/j.compstruct.2017.08.086

Afzal, S., Shahzada, K., Fahad, M., Saeed, S., & Ashraf, M. (2014). Assessment of early-age autogenous shrinkage strains in concrete using bentonite clay as internal curing technique. *Construction and Building Materials, 66*, 403–409. https://doi.org/10.1016/j.conbuildmat.2014.05.051

Ait Mohamed Amer, A., Ezziane, K., Bougara, A., & Adjoudj, M. (2016). Rheological and mechanical behavior of concrete made with pre-saturated and dried recycled concrete aggregates. *Construction and Building Materials, 123*, 300–308. https://doi.org/10.1016/j.conbuildmat.2016.06.107

Akbar, J., Alam, B., Ashraf, M., Afzal, S., Ahmad, A., & Shahzada, K. (January 2013). Evaluating the effect of bentonite on strength and durability of high performance concrete. *International Journal of Advanced Structures and Geotechnical Engineering, 2*(1), 1–5.

Akbarnezhad, A., Ong, K. C. G., Zhang, M. H., Tam, C. T., & Foo, T. W. J. (2011). Microwave assisted beneficiation of recycled concrete aggregates. *Construction and Building Materials, 25*(8), 3469–3479. https://doi.org/10.1016/j.conbuildmat.2011.03.038

Ann, K. Y., Moon, H. Y., Kim, Y. B., & Ryou, J. (2008). Durability of recycled aggregate concrete using pozzolanic materials. *Waste Management, 28*(6), 993–999. https://doi.org/10.1016/j.wasman.2007.03.003

Arredondo-Rea, S. P., Corral-Higuera, R., Gómez-Soberón, J. M., Castorena-González, J. H., Orozco-Carmona, V., & Almaral-Sánchez, J. L. (2012). Carbonation rate and reinforcing steel corrosion of concretes with recycled concrete aggregates and supplementary cementing materials. *International Journal of Electrochemical Science, 7*(2), 1602–1610.

Brauers, W. K. M., Zavadskas, Edmundas K., Peldschus, F., & Turskis, Z. (2008). Multi-objective decision-making for road design. *Transport, 23*(3), 183–193, doi: 10.3846/1648 4142.2008.23.183-193.

Chen, C. H., Huang, R., Wu, J. K., & Yang, C. C. (2006). Waste E-glass particles used in cementitious mixtures. *Cement and Concrete Research, 36*(3), 449–456. https://doi.org/10.1016/j.cemconres.2005.12.010

Chen, G., Lee, H., Young, K. L., Yue, P. L., Wong, A., Tao, T., & Choi, K. K. (2002). Glass recycling in cement production—An innovative approach. *Waste Management, 22*(7), 747–753. https://doi.org/10.1016/s0956-053x(02)00047-8

Corinaldesi, V., Gnappi, G., Moriconi, G., & Montenero, A. (2005). Reuse of ground waste glass as aggregate for mortars. *Waste Management, 25*(2), 197–201. https://doi.org/10.1016/j.wasman.2004.12.009

Dhivyana, R. (2015). An experimental study on concrete using bentonite and steel slag. In *National Conference on Research Advances in Communication, Computation, Electrical Science and Structures (NCRACCESS-2015)*, ISSN: 2348–8352.

Hamemrnik, J. D., & Frantz, G. C. (1991). Physical and chemical properties of municipal solid waste fly ash. *ACI Materials Journal, 88*(3), 294–301.

Hwang, C. L., & Yoon, K. (1981). *Multiple attribute decision making methods and applications*, Springer, Berlin Heidelberg.

Ismail, Z. Z., & Al-Hashmi, E. A. (2009). Recycling of waste glass as a partial replacement for fine aggregate in concrete. *Waste Management, 29*(2), 655–659. https://doi.org/10.1016/j.wasman.2008.08.012

Karande, P., & Chakraborty, S. (2012). Application of multi-objective optimization on the basis of ratio analysis (MOORA) method for materials selection. *Materials and Design, 37*, 317–324.

Karthikeyan, M., Ramachandran, R. P., Nandhini, A., & Vinodha, R. (2015). Application on partial substitute of cement by bentonite in concrete. *International Journal of Chem Tech Research, 8*(11), 384–388.

Kou, G., Lu, Y., Peng, Y., & Shi, Y. (2012). Evaluation of classification algorithms using MCDM and rank correlation. *International Journal of Information Technology & Decision Making, 11*(1), 197–225. https://doi.org/10.1142/S0219622012500095

Madhavi, T. P., Sampathkumar, V., & Gunasekaran, P. (2013). Partial replacement of cement and fine aggregate by using fly ash and glass aggregate. *International Journal of Research in Engineering and Technology, 2*(13), 351–355. ISSN: 2321-7308

Memon, S. A., Arsalan, R., Khan, S., & Lo, T. Y. (2012). Utilization of Pakistani bentonite as partial replacement of cement in concrete. *Construction and Building Materials, 30*, 237–242. https://doi.org/10.1016/j.conbuildmat.2011.11.021

Metwally, I. M. (2007). Investigations on the performance of concrete made with blended finely milled waste glass. *Advances in Structural Engineering, 10*(1), 47–53. https://doi.org/10.1260/136943307780150823

Mirza, J., Riaz, M., Naseer, A., Rehman, F., Khan, A. N., & Ali, Q. (2009). Pakistani bentonite in mortars and concrete as low cost construction material. *Applied Clay Science, 45*(4), 220–226. https://doi.org/10.1016/j.clay.2009.06.011

Opricovic, S. (1998). *Multi-criteria optimization of civil engineering systems*, Faculty of Civil Engineering, Belgrade.

Park, S. B., Lee, B. C., & Kim, J. H. (2004). Studies on mechanical properties of concrete containing waste glass aggregate. *Cement and Concrete Research, 34*(12), 2181–2189. https://doi.org/10.1016/j.cemconres.2004.02.006

Pavlicic, D. (2001). Normalization affects the results of MADM methods. *Yugoslav Journal of Operations Research (YUJOR), 11*(2), 251–265.

Rakshvir, M., & Barai, S. V. (2006). Studies on recycled aggregates-based concrete. *Waste Management and Research, 24*(3), 225–233. https://doi.org/10.1177/0734242X06064820

Shayan, A., & Xu, A. (2004). Value-added utilization of waste glass in concrete. *Cement and Concrete Research, 34*(1), 81–89. https://doi.org/10.1016/S0008-8846(03)00251-5

Shayan, A., & Xu, A. (2006). Performance of glass powder as a pozzolanic material in concrete: A field trial on concrete slabs. *Cement and Concrete Research, 36*(3), 457–468. https://doi.org/10.1016/j.cemconres.2005.12.012

Shekhawat, B. S., & Aggarwal, V. (2014). Utilization of waste glass powder in concrete. *International Journal of Innovative Research in Science, Engineering and Technology, 3*(7), 14822–14826.

Sobolev, K., Türker, P., Soboleva, S., & Iscioglu, G. (2007). Utilization of waste glass in ECO-cement: Strength properties and microstructural observations. *Waste Management, 27*(7), 971–976. https://doi.org/10.1016/j.wasman.2006.07.014

Stanujkic, D., Magdalinovic, N., Stojanovic, S., & Jovanovic, R. (2012). Extension of ratio system part of MOORA method for solving decision-making problems with interval data. *Informatica, 23*(1), 141–154.

Topçu, İ. B., & Canbaz, M. (2004). Properties of concrete containing waste glass. *Cement and Concrete Research, 34*(2), 267–274. https://doi.org/10.1016/j.cemconres.2003.07.003

Triantaphyllou, E. (2000). *Multi-criteria decision making methods: A comparative study.* Kluwer Academic Publishers, Dordrecht.

Ugur, L. O., Yuksel, E., & Erdal, M. (2018). Selection of reinforced concrete formwork system with MOORA multi criteria decision making method. In *International Conference on Engineering, Technology and Innovation*, Budapest.

Vasudevan, G., & Pillay, K. G. S. (2013). Performance of using waste glass powder in concrete as replacement of cement. *American Journal of Engineering Research, 2*(12), 175–181.

Verian, K. P., Ashraf, W., & Cao, Y. (2018). Properties of recycled concrete aggregate and their influence in new concrete production. *Resources, Conservation & Recycling, 133*, 30–49. https://doi.org/10.1016/j.resconrec.2018.02.005.

Vijayakumar, G., Vishaliny, H., & Govindarajulu, D. (2013). Studies on glass powder as partial replacement of cement in concrete production. *International Journal of Emerging Technology and Advanced Engineering, 3*(2), 1–12.

Chapter 6

Green energy harvesting using high entropy thermoelectric alloys

Arun Raphel

National Institute of Technology, Tiruchirappalli, Tamil Nadu, India

Viswajyothi College of Engineering and Technology, Ernakulam, India

P. Vivekanandhan

International Advanced Research Centre for Powder Metallurgy and New Materials, Chennai, Tamil Nadu, India

National Institute of Technology, Tiruchirappalli, Tamil Nadu, India

S. Kumaran

National Institute of Technology, Tiruchirappalli, Tamil Nadu, India

CONTENTS

DOI: 10.1201/9781003257714-6

6.1 INTRODUCTION

Globally, the escalation in the demand for energy and the increased aware-ness of greenhouse effects has become an intense threat to human beings in recent decades. The exponentially increasing energy requirement, increased global population, environmental impact from the combustion of fossil fuels, and a dramatic decline in the non-renewable energy resources have created serious imbalances in nature and society. According to the global energy and CO_2 status report for 2019, the global demand for electric-ity has been accelerating since 2010. In 2018, for example, there was a 4% increase in electricity demand. As per Renewables 2021 Global Status Report (Figure 6.1), there was a significant increase in the total final energy

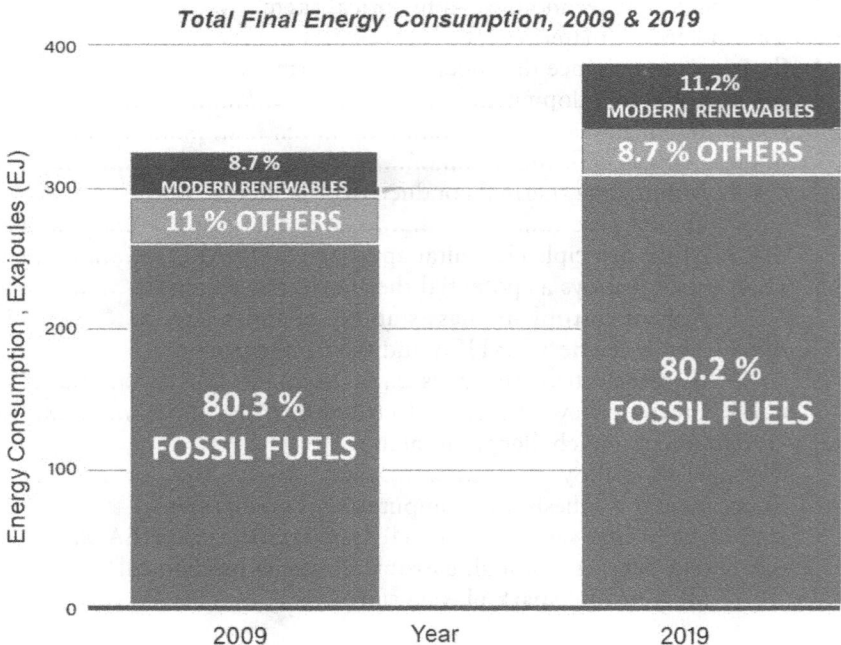

Figure 6.1 Comparison of total final energy consumption for the years 2009 & 2019. (Renewables 2021 Global Status Report).

consumption (TFEC). Even though there was a noticeable rise in the energy consumption via modern renewable resources, from 8.7% in 2009 to 11.2% in 2019, the TFEC contribution of fossil energy reserve (80.2%) has remains stable over a decade. This is a serious concern as a quantitative measure of the fossil energy consumption is much higher due to increased energy requirement despite of the constant TFEC contribution due to increased energy requirements.

Thus, the necessity to develop alternative energy sources over conventional fossil fuel is now quite clear, even though it may be challenging. In this context, intense research has progressed over the years to develop potential sustainable and highly efficient renewable energy technologies. Even though these non-renewable resources could compensate the energy requirement to a greater degree, they cannot completely replace conventional fossil fuel energy resources. With regard to the statistical data, of the total amounted of wasted energy worldwide, more than 60% is accounted as thermal energy in the form of heat. This includes heat loss from the human body, kitchens, automobiles, and industries. Thus, even if we could recover a small percentage of this wasted heat energy, it could create a promising improvement in the energy efficiency and sustainability of the world.

In many countries, such as Germany, electricity generation from renewable sources now surpasses the generation from coal reserves. However, in terms of the increase in global generated electricity from single sources, coal remains the largest contributor, followed by natural gas. The coal itself accounts for a rise of 2.5% in CO_2 emissions. So, the dream of a complete switch over from non-renewable to renewable green energy technology may take many more years. Accordingly, significant efforts should be made to minimize the utilization of non-renewable reserves and thereby reduce undesirable emissions. Scavenging the wasted heat can enhance the net energy conversion efficiency of the non-renewable energy sources and reduce fossil reserve consumption. Thermoelectric generators can achieve this by converting heat energy into electric energy and vice versa.

Thermoelectric materials are defined as solid-state semiconductors that directly convert thermal energy into electrical energy pronounced by charge carriers. The performance efficiency of thermoelectric materials is measured using a dimensionless quantity termed the thermoelectric figure of merit (ZT), and which is defined as

$$ZT = \frac{S^2\sigma}{k} \tag{6.1}$$

where σ is the electrical conductivity, S is the Seebeck coefficient, and k is the thermal conductivity of the corresponding material. Thus, in order to have a high ZT, a material should possess a high S and σ with minimal value for k. High electrical conductivity with minimum joule heating will increase the generated thermoelectric voltage. By contrast, the lower value of thermal conductivity will contain the heat within the material, reduce the heat transfer between hot and cold junctions, and retain the temperature gradient. In addition to the simplicity in terms of energy conversion, it is further conceived by various appreciable factors such as long service span, noiseless operation, no fueling, no outlet of carbon footprints, economically cheaper, scalable, zero maintenance cost, etc. Although several renewable technologies are currently being appreciated, the direct and reliable conversion of waste heat into electricity with no carbon footprints has made thermoelectric technology both attractive and promising. HEA can achieve good electrical conductivity due to good crystal structure symmetry and low thermal conductivity due to severe lattice distortion and phonon scattering. Furthermore, the elemental doping with a tuned electronic band structure can optimize the thermoelectric performance of HEA.

This chapter details the role of high entropy phenomena in waste heat recovery via thermoelectric generators and recent advancements in HEA for thermoelectric applications. Further, the chapter also provides a comprehensive overview on design aspects and material selection in HEA, synthesis routes, thermodynamic parameters and thermoelectric properties, role of DFT studies in understanding the underlying principles of thermoelectric HEA.

6.2 THERMOELECTRIC TECHNOLOGY—A HISTORICAL SKETCH

In 1821, Thomas Johann Seebeck discovered the Seebeck effect. His research showed that an applied temperature difference in a material can cause charge carrier (electrons and holes) diffusion along the temperature gradient, resulting in a current flow in the material. Jean Charles Athanase Peltier, a French physicist, demonstrated later in 1834 that this Seebeck effect could be reversed. He found that current flow through a material can result in heat absorption at one end and heat rejection at the other end. This heat absorption can create active cooling, and this phenomenon is termed the Peltier effect. These two discoveries laid the foundation for the development of thermoelectric generators and the later Peltier cooler. In 1854, William Thomson established an excellent explanation for the relationship between the Seebeck and Peltier effects. Following this, there were no significant advancements in thermoelectric research until 1909, when Edmund Altenkirch derived the maximum possible efficiency of a thermoelectric generator using the constant property model. In his book *Semiconductor Thermoelements and Thermoelectric Cooling* (1949), Abram Fedorovich Ioffe introduced the figure of merit (ZT) concept and also formulated the modern theory of thermoelectrics.

Intense research in thermoelectrics began after the invention of radioisotope thermoelectric generators (RTGs) for space application in 1969. Under this process, a significant temperature gradient is established between the heat generated from the radioisotope fuel and the low-temperature space atmosphere, which is desirable for thermoelectric power generation. That is the reason why, even after a combined operating time of over 300 years, not even a single thermocouple has failed to function. NASA widely used RTGs for various space missions in the period 1969 from 2000. Even though they have only exhibited limited power generation capability, thermoelectric generators became an integral part of space missions due to their high reliability and maintenance-free operation. From the beginning of the 21st century to date, comprehensive research is becoming widespread in thermoelectrics, using advanced concepts such as nanostructuring, band engineering, and the phonon glass electron crystal (PGEC) concept, to enhance thermoelectric power generation efficiency.

However, there is a substantial hindrance in achieving the higher efficiency of thermoelectric generators (TEGs). The current efficiency of the TEG systems is far less than 10%. The major reasons for this include limited potential materials, counteracting interdependence over the material properties, and unavoidable contact resistance generated during device fabrication. Thermoelectric technology development has shown a commendable phase shift over the past couple of years due to the development of a broad spectrum of new-generation materials with novel chemistry and contemporary processing routes. Advancements in nanotechnology, which include nanostructured material developments, enhancing the thermoelectric properties using nano inclusions, developing thermoelectric nanocomposites, nano-scale device fabrications, etc., have contributed significantly to the enhancement of thermoelectric performance. Other viable strategies, such as bang engineering, entropy engineering, and resonant doping, could fine-tune the thermoelectric material properties to a greater extent. Another critical factor that governs the energy conversion materials is thermoelectric compatibility. In a thermoelectric module, it is possible to connect many thermoelectric pairs electrically in series to obtain the required electricity. The materials used should offer minimum thermal resistance in the contact parts of the circuit to minimize the energy loss within the device. In addition, another important factor involves the limitations in the operating temperature range of the thermoelectric materials. Different materials exhibit their optimum thermoelectric performance in different temperature regimes. Certain materials, such as Bi_2Te_3, Sb_3Te_3, etc., show excellent thermoelectric properties at near-room temperature but have proven unsuitable for higher temperatures. Promising materials, such as PbTe, SnTe, etc., are suitable for mid-temperature thermoelectric applications, whereas SiGe, Mg_2Si are high-temperature thermoelectric materials. Thermoelectric generators with similar material types were designed for a specific operating temperature range in earlier days. However, with the advanced material development and fabrications route, it is now possible to make a thermoelectric generator module incorporated with different material segments designed for different temperature ranges. Materials used in such generators possess excellent thermoelectric compatibility and minimum contact resistance, thereby increasing the module's overall effectiveness.

6.3 THERMOELECTRIC MATERIALS

The thermoelectric materials can be generally classified based on the operating temperature range and the time they were developed. Based according to temperature, they can be classified into low-temperature (<400 K), mid-temperature (500 K–900 K), and high-temperature (>900 K) thermoelectric materials (Zhang and Zhao 2015). Bi_2Te_3, Sb_2Te_3, and Bi_2Se_3 are all proven low-temperature thermoelectric materials for near-room temperature applications. Chalcogenides-based solid solutions, such as PbTe, PbSe, PbS (lead-based), and SnTe, SnSe, and SnS (tin-based), are more suitable for mid-temperature applications (Meng et al. 2020; Gayner et al. 2018; Qin et al. 2019; Zhao et al. 2019, Zhao et al. 2016; Wang et al. 2019). Si–Ge alloys were mainly used as high-temperature thermoelectric materials above 900 K. To minimize inherent losses during device fabrication, both p-type and n-type materials could be synthesized from the same Si–Ge material chemistry. Apart from these classifications, attempts were made to create thermoelectric materials for active cooling below room temperature. Thermoelectric generators operating below room temperature (Bi, Sb)$_2$(Te, Se)$_3$ were used as p-type materials for such ultra-low temperatures, and the BiSb alloy was used as an n-type material. However, undesirable mechanical properties of BiSb alloys limit the device fabrication (Snyder and Toberer 2008).

Developments in thermoelectric material research can be classified into three generations. Following the introduction of the Seebeck effect in 1921 by Thomas Johann Seebeck, there were no significant developments for nearly a decade. In 1929, Edmund Altenkirch formulated a property model to derive the maximum possible efficiency of a thermoelectric generator. Later, in 1949, Abram Fedorovich Ioffe introduced the concept of the figure of merit ZT, thereby leading to the development of modern thermoelectric research. The first-generation thermoelectric materials, such as Bi_2Te_3, PbTe, and SiGe, possessed a maximum ZT of 1. The maximum conversion efficiency for devices made of first-generation materials was estimated to be less than 5%. The thermoelectric research after 1990 mainly focused on size effect in improving thermoelectric properties via nanostructuring, nano-scale inclusions, synthesis of nanomaterial, etc. These materials could be categorized as second-generation thermoelectric materials. A maximum ZT of 1.7 and an energy conversion efficiency rate of 11% to 15% were achieved by adopting a size effect. At a later age, intense research was carried out in thermoelectric materials employing various strategies like band structure modification, hierarchical architecture, band convergence, entropy engineering, etc. This period of advanced thermoelectric research can be considered the third generation of thermoelectric material development. Many novel materials were synthesized during this period, like Cu_2Se, half-Heusler alloys, high entropy alloys, etc., with promising thermoelectric performance.

6.4 STRATEGIES TO ENHANCE THERMOELECTRIC PERFORMANCE

6.4.1 Elemental doping

Dopant plays a crucial role in controlling charge carrier (holes and electrons), the nature of semi-conduction (p-type or n-type semi conduction), and the electron density of states of the material. Doping determines the carrier concentration and the fermi-level. In general, increasing the carrier concentration via doping will compromise the Seebeck coefficient. Certain dopants can converge conduction or valance band higher degree of valley degeneracy. Thus, the careful selection of the dopants in optimal concentration is required for fine-tuning the band structure.

6.4.2 Band engineering

Modifying electronic band structure such as increased bandgap, the distortion of the electron density of states, and increasing the effective mass can contribute to notable enhancement in the absolute Seebeck coefficient and thereby the thermoelectric figure of merit of the material. Introducing thallium (Tl) as resonant doping in PbTe significantly alters the band structure by DOS distortion and increased effective mass, resulting in a prominent ZT of 1.5 at 773 K (Heremans et al. 2008). The alloying of elements, such as magnesium (Mg) or manganese (Mn), to form $Pb_{(1-x)}M_xTe$ solid solution, where M is the solute factor (M = Mg/Mn), results in a remarkable variation in the band structure. This is due to the combined effect of increased bandgap resulting from the downward shifting of valance band and band convergence between a light hole band (L band) and a heavy hole band (Σ band) (Zhang and Zhao 2015). Thus, the band structure modification via resonant state doping and the convergence of valence bands resulted in a significant improvement in the Seebeck coefficient.

6.4.3 Nanostructuring

Band engineering and doping in thermoelectric materials help in improving the thermoelectric power factor through increasing either electrical conductivity or the Seebeck coefficient. Nevertheless, to achieve higher ZT, a thermoelectric material should possess a lower value for thermal conductivity. In general, increasing electrical conductivity will increase the electronic contribution of thermal conductivity. Thus, it is vital to restrict the total thermal conductivity to the desirable range; otherwise the ZT will decrease despite the increased power factor. One effective way of reducing the thermal conductivity is by reducing the lattice contribution of thermal conductivity. This could be achieved in many ways, such as enhancing phonon scattering via multi-phase composites, nano-scale inclusions, nano structuring, etc. When the material's grain size is reduced to the nanometre range, a greater number of nanostructured surfaces/interfaces will be available for phonon scattering. This could effectively contribute towards reducing the lattice thermal conductivity. In general, nanostructuring shows an inferior effect over the thermoelectric power factor.

6.4.4 Minimizing thermal conductivity via hierarchical architecture

To reduce the total thermal conductivity, it is advisable to reduce lattice thermal conductivity regardless of charge carrier concentration. On the other hand, electronic thermal conductivity and electrical conductivity are influenced by carrier concentration. The mean free path of phonons determines the thermal conductivity of the lattice. In contrast to electrons that travel with a narrow range of wavelengths, phonons propagate across a broad spectrum of wavelengths. Thus, instead of a single scattering mechanism, different strategies must be employed for the effective scattering of such short- (1–10 nm), medium- (10–100 nm), and long-wavelength (100–1000 nm) phonons. Figure 6.2 illustrates a schematic representation of phonon scattering via a combination of point defects, the formation of nano-precipitates, and grain boundary scattering by nanostructuring. Alloying/doping can aid in point defect scattering; nano-inclusions can scatter mid-wavelength phonons, whereas grain boundaries can scatter long-wavelength phonons. By introducing such a multi-scale strategies lattice, thermal conductivity can be significantly lowered to its theoretical limit.

6.4.5 Multi-principle elemental approach (MPEA)

MPEA are new alloy design concepts generated to tailor materials for various structural and functional applications. The principle of MPEA can also be applied to enhance different thermoelectric properties. The severe lattice distortion is resulting from the complex chemistry of MPEA aid in hindering lattice vibrations and reducing the thermal conductivity. Also, large lattice strains favor the formation of multiple-phase boundaries to lower lattice thermal conductivity. Pseudo-binary compounds of Te-Sb-Ge-Ag (commonly called TAGS) is a classic example of MPEA with complex chemistry (Perumal et al. 2016). Increasing the composition complexity further

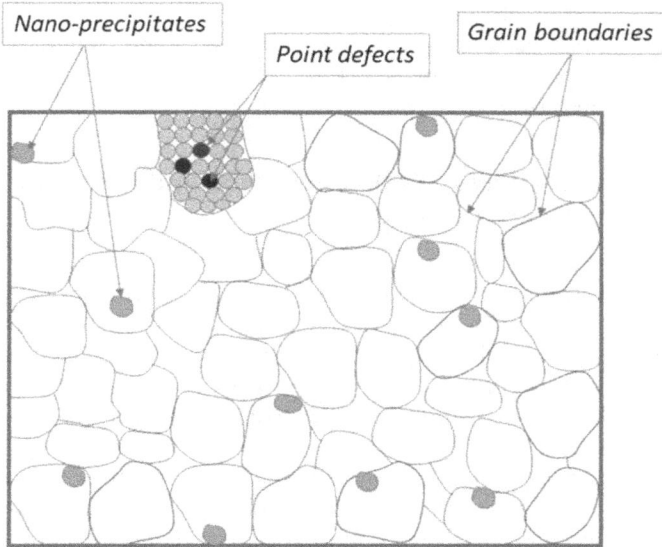

Figure 6.2 Multi-scale strategies for scattering phonons of different wavelength.

reduces the thermal conductivity. When compared with binary and ternary alloy systems, multi-principle elemental HH alloys possess lower thermal conductivity and higher ZT (Shen et al. 2001).

6.5 HIGH ENTROPY ALLOYS AS POTENTIAL THERMOELECTRIC MATERIALS

6.5.1 Role of entropy in phase stability of alloys

The phase stability of HEA is governed by the Gibbs free energy, i.e., $\Delta G_{mix} = \Delta H_{mix} - T\Delta S_{mix}$. Gibbs free energy can be lowered either by increasing the enthalpy of mixing or by increasing the entropy of mixing. In the case of conventional alloys, enthalpy of mixing (ΔH_{mix}) is the governing factor for alloy phase stability. When the ΔH_{mix} value is more negative, ΔG_{mix} is significantly lowered, and the alloy becomes more stable. In the case of high entropy alloys, it is the ΔS_{mix} which is the governing factor when compared to the ΔH_{mix}. The ΔS_{mix} consists of configurational entropy and non-configurational entropy. As ΔS_{mix} is made more positive, ΔG_{mix} becomes more negative, and the stability of the alloy increases. The configurational entropy of the HEA increases when the system's constituent elements increase, and their distribution is more random. The non-configurational entropy has two main contributions, i.e., vibrational entropy and electronic entropy. Electronic entropy has no considerable effect on alloy phase stability, whereas vibrational entropy may either positively or negatively influence alloy phase stability. Hence, in HEA, configurational entropy dominates over electronic and vibrational entropy in terms of stabilizing the alloy. That is why configurational entropy is directly referred to as the entropy of mixing in certain literature.

6.5.2 The core effects of HEA and their influence on thermoelectric properties

A wide range of thermoelectric material properties is affected by the core effects of HEA, such as the high entropy effect, the severe lattice distortion, and the slow diffusion of atoms (Raphel et al., 2020). In general, increasing the number of constituent elements of a thermoelectric material should boost the configurational entropy and Seebeck coefficient via the formation of new bands or band convergence. However, the electronic properties purely depend on how the added elements via doping/alloying affect the carrier concentration and mobility of the charge carrier. In certain materials, charge carrier concentration and mobility simultaneously increase, whereas in certain other materials, even though carrier concentration increases, mobility decreases. This is the result of dopants/alloying elements acting as point defect scattering centers that shorten the mean effective path of electrons. An increasing number of elements will reduce the thermal conductivity due to increased lattice distortion and phonon scattering centers. Thus, if carefully controlled, the inherent core effects of HEA can have a positive impact on the thermoelectric performance of the material.

6.5.3 Entropic view of thermoelectric properties

The configurational entropy shows a direct dependence over the lattice thermal conductivity (k_l) and inverse dependence over the mean free path of the charge carrier electrons (μ). Increasing the configurational entropy enhances the deformation potential resulting from the lattice distortion. This lattice distortion will increase with increased constituent elements, atomic size difference, and atomic mass difference of the elements present. All of this will significantly contribute to the scattering of heat-carrying phonons. Hence the high entropy effect and severe lattice distortion effects, which are two inherent core effects of HEA, can help in reducing lattice thermal conductivity and enhancing the ZT. However, an increase in configurational entropy will increase the electrical resistance (ρ), which reduces the electrical conductivity.

Another prominent effect of configurational entropy is that it can offset the stain energy, resulting in a more stabilized single-phase solid solution. If the atomic size difference among the constituent elements increases, more elements are required in the HEA to form a single-phase reliable solution. The entropy of the alloy also influences thermopower or the Seebeck effect. As entropy per charge increases, the Seebeck coefficient also increases.

Configurational entropy can increase the crystal structure symmetry and improve the electron density of the state's effective mass, which can significantly alter the electron band structure of the material. This band structure modification via the formation of an additional band or band convergence can result in improved Seebeck coefficient and thermoelectric power factor.

6.6 DEVELOPMENT AND CHALLENGES IN HIGH ENTROPY THERMOELECTRIC ALLOYS

Contrary to conventional alloy systems, high entropy alloys possess complex material chemistry due to four or more principal elements. The multi-component constituent elements may significantly differ in terms of atomic size, mass, and melting point. Thus, the synthesis of such alloys requires extreme care to achieve the desired phase characteristics. The processing route plays a crucial role in the HEA, especially for the achievement of superior functional properties such as higher thermoelectric energy conversion efficiency. The first attempt to investigate the thermoelectric performance of a high entropy alloy was carried on $Al_xCoCrFeNi$ where x = 0 to 0.3 (Shafeie et al. 2015). Here alloys were synthesized via an arc melting route in a titanium-gettered argon environment. The alloy formed was re-melted again using the arc melting furnace five times to maintain chemical homogeneity. A similar processing route via arc melting was also used for the synthesis of other thermoelectric HEAs (Dong et al. 2018; Han et al. 2020).

Thermoelectric HEAs of PbSnTeSe and $BiSbTe_{1.5}Se_{1.5}$ were developed using a specialized processing route consisting of melting, quenching, annealing, and spark plasma sintering (Fan et al. 2016; Fan et al. 2017). This consists of a prolonged processing duration involved with various complex stages. It requires heating from room temperature to 1373 K at 1 K/m (a period of approximately18 hrs), and then subsequently soaking at 1373 K for 6 hrs. Then it was involved with water quenching and annealing at 873 K for 120 hrs, followed by grinding to powders and being consolidated into bulk using SPS at 50 MPa. Hence, the past work emphasizes the synthesis procedure, which demands 144 hrs of processing time for PbSnTeSe HEA. A recent study to synthesize $Sn_{0.25}Pb_{0.25}Mn_{0.25}Ge_{0.25}Te$ thermoelectric HEA based on melting and annealing followed by SPS consolidation involves a considerable processing duration of more than ten days. The high processing time and careful control over the multi-component material chemistry were always challenging parts of the synthesizing process of thermoelectric HEAs.

A powder metallurgical process combining mechanical alloying (MA) and spark plasma sintering (SPS) might be viable for the production of thermoelectric HEA. MA and SPS were effective solid-state synthesis routes for developing high entropy alloys with better process control. Further, they have also proven synthesis routes for conventional thermoelectric material systems, such as Bi_2Te_3, PbTe SiGe, etc. The mechanical alloying followed by spark plasma sintering is a potential route to synthesize thermoelectric high entropy alloys. They could produce the required alloy phase with optimized process parameters in a concise processing duration. Also, since there is no melting of constituent elements, this solid-state synthesis route ensures better control over the material chemistry than the conventional melt route synthesis. Further lattice distortion induced during MA, and the ability to form nanocrystalline materials, help in effective phonon scattering, reducing the lattice thermal conductivity to a greater extent. This can help lower total thermal conductivity and improve the thermoelectric figure of merit. Hence, an optimized synthesis technique combining mechanical alloying and spark plasma sintering can be a potential synthesis route for developing future-generation HEAs with superior thermoelectric performance.

6.7 EXPERIMENTAL SYNTHESIS AND COMPUTATIONAL PREDICTIONS IN LEAD-TIN-TELLURIUM-SELENIUM (PBSNTESE) THERMOELECTRIC HEA

6.7.1 The synthesis of high entropy alloys via mechanical alloying and spark plasma sintering

As described in the preceding section, the development of multi-component alloys such as high entropy alloys through a powder metallurgical approach combining mechanical alloying and spark plasma sintering offers improved process control, uniform elemental distribution, and better homogeneity. All the elemental powders of lead, tin, tellurium, and selenium were taken in the equimolar concentration of 25% each to synthesize PbSnTeSe HEA. The selected powders with dopants of suitable atomic concentration were milled in a high-energy planetary mill. Mechanical alloying (MA) was carried out for an optimized time of 5 hr with a ball-to-powder ratio (BPR) of 15:1 and a milling speed of 300 rpm. The powder filling was carried out in a high-purity inert argon atmosphere. After MA, the alloy power was used to fill the cylindrical graphite die assembly and these were filled in high-density graphite die of sample dimension 12 mm diameter and 6 mm height under vacuum atmosphere (<10 Pa). The powder compaction was carried out using the spark plasma sintering (SPS) technique. The single-step sintering was performed from room temperature (RT) to 325°C at a pressure of 50 MPa and a heating rate of 50°C per minute. As soon as the sample reached 325°C, it was soaked for 5 minutes before being cooled to room temperature in the SPS furnace. The systematic representation of solid-state synthesis and densification of PbSnTeSe HEA is shown in Figure 6.3.

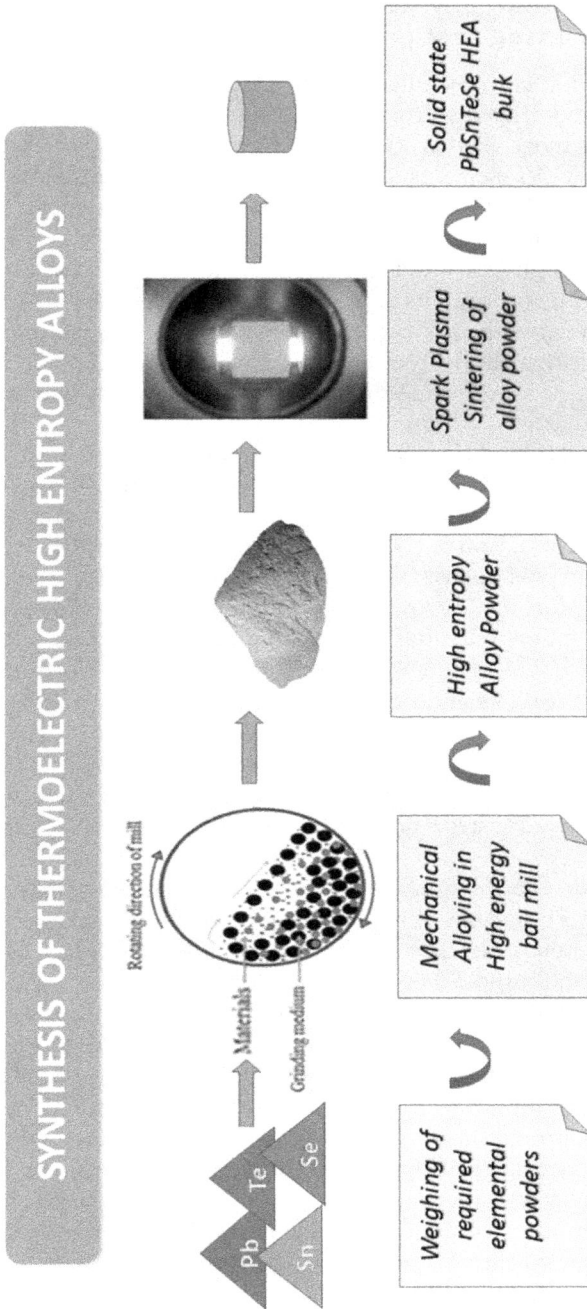

SYNTHESIS OF THERMOELECTRIC HIGH ENTROPY ALLOYS

Rotating direction of mill

Materials

Grinding medium

Pb   Te

Sn   Se

Weighing of required elemental powders

Mechanical Alloying in High energy ball mill

High entropy Alloy Powder

Spark Plasma Sintering of alloy powder

Solid state PbSnTeSe HEA bulk

Figure 6.3 Systematic representation of the synthesis of PnSnTeSe HEA using MA and SPS.

6.7.2 Determination of crystal structure of high entropy alloys by density functional theory (DFT) based first principle approach

Engineering high entropy alloys with multi-component elements often results in severe lattice distortion, impacting phonon scattering and lattice thermal conductivity. In this study, the quaternary alloy PbSnTeSe formed by equimolar IV–VI elements shows significantly low lattice thermal conductivity of about 0.45 $Wm^{-1}K^{-1}$ at a lower mid-range temperature (450–625 K). Although it exhibits low lattice thermal conductivity, the ZT value is not relatively high (ZT = 0.47) (Raphel et al. 2021a). Here lies room for optimizing the HEA to reap its inherent structural benefits. Na was added to the PbSnTeSe alloy with various doping concentrations to fine-tune the thermoelectric performance (Raphel et al. 2022). This would potentially make PbSnTeSe HEA be a promising thermoelectric material even at the medium temperature range.

The spin unrestricted all-electron density functional theory (DFT) calculations were performed for the PbSnTeSe high entropy alloys doped with Na using density functional calculations on molecules and the three-dimensional periodic solids (DMol³) package (Delley 1990; Delley 2000). Double numeric plus polarization (DND) basis set with the fine k-point set was employed for the electronic calculations of the HEA system. The exchange-correlation potential was calculated using generalized gradient approximation (GGA) Perdew-Burke-Ernzerhof (PBE) (Heyd et al. 2003; Perdew et al. 1996). The undoped HEA composition was considered to have a face-centered cubic (FCC) crystal structure. Virtual crystal approximation (VCA) was employed to reduce the need for many supercells and associated computational costs. It becomes imperative to note that VCA works quite well with disordered systems and simultaneously reduces the computational time.

The PbSnTeSe HEA exhibits a NaCl-type FCC structure based on DFT first principles. Pb and Sn atoms are equally likely to occupy the corners and faces of cubic unit cells (with an occupational probability of 50% each). Atoms of Te and Se possess equal probabilities of occupying edge centers. The virtual crystal approximation (VCA) technique has been used to model the crystal structure for $Pb_{0.99}SnTeSe-Na_{0.01}$ HEA (Figure 6.4b). The blue-colored atom represents the effect of the addition of Na, which is composed of 48%Pb-50%Sn-2%Na. Since Pb and Sn sites possess equal site occupation probability, it is difficult to estimate the exact doping location. Further, arguing the dopant site probability between the Pb and Sn sites is difficult because it is a HEA system with five elements. To date, there is no experimental or computational evidence showing the site preference of dopant Na atom into Pb or Sn. The computational model used in this work is based on the principle of virtual crystal approximation (VCA) for the same reason. This technique is used because the dopant site is not

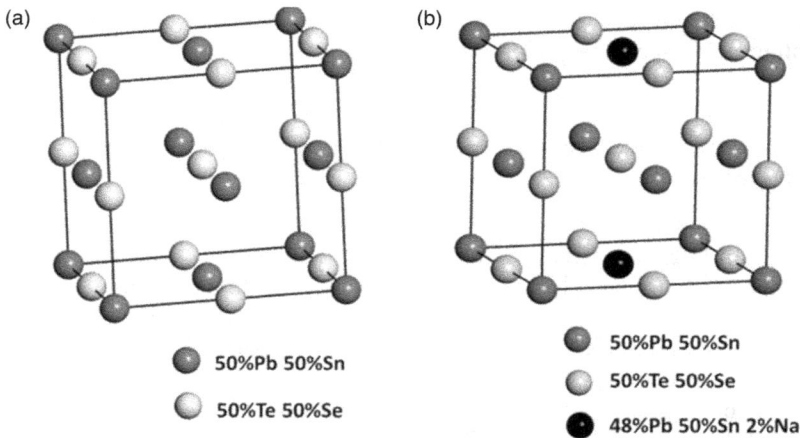

Figure 6.4 Estimated crystal structure of (a) PbSnTeSe HEA and (b) $Pb_{0.99}SnTeSe$-$Na_{0.01}$ HEA.

entirely known. The VCA method works by creating mixture atoms, wherein the atom compositions are altered to represent the HEA system. Therefore, theoretically, the dopant Na is almost equally likely to be in either the Pb or the Sn site as per the computational setup.

Nevertheless, it is quite interesting to explore any clues to understanding the site occupancy of the dopant (between Pb and Sn atoms). So, we performed density functional theory (DFT) total energy calculations with GGA-PBE approximation. The Na dopant was introduced using the VCA technique independently on both Pb and Sn sites. The resulting total energy of the relaxed system of the base system (no dopant), the Na dopant at the Pb site, and the Na dopant at the Sn site are summarized in Table 6.1. It is to be noted that there is no change in total energy values calculated for the base and the system with the Na dopant at the Pb site. In contrast, there is a significantly lower total energy in the system with the Na dopant at the Sn system. As a result, we conclude that doping at both sites (Pb/Sn) is viable, given that the system turns out to be stable, and the total energy is equal to, or lower than, the parent system. Interestingly, the Na dopant is likely to

Table 6.1 The total energy of relaxed structure

HEA System	Total energy (Ha)
PbSnTeSe	−33705.314982
$Pb_{.99}SnTeSe$-$Na_{.01}$	−33705.314982
$PbSn_{.99}TeSe$-$Na_{.01}$	−47208.169058

take the Sn site rather than the Pb site, as the total energy of the resulting configuration seems to be lower.

6.7.3 Predicting the band structure and electron density of states of high entropy alloys via first principle approach

The band structure and electron density of states play a crucial role in understanding the transportation of charge carriers and the semiconducting behavior of the alloy systems. However, the first-principle studies of multi-component systems such as high entropy alloys are rare. The first-ever attempt to simulate the band structure and DOS of Na doped PbSnTeSe HEA shows encouraging results (Raphel et al. 2022). A significant enhancement in band gap, from 0.327 eV to 0.490 eV, is obtained for $Pb_{0.99}SnTeSe$-$Na_{0.01}$ HEA compared to PbSnTeSe HEA (Figure 6.5a and c). A significant increase in the absolute Seebeck coefficient and ZT could result from this. Further Na doping in PbSnTeSe HEA created a notable increase in the electron DOS. Based on the DOS, Na addition helps fine-tune the peaks, at least to a certain extent, in $Pb_{0.99}SnTeSe$-$Na_{0.01}$ HEA compared to PbSnTeSe HEA (Figure 6.5b and d). Unlike the initially merged peaks, more pronounced peaks are seen in $Pb_{0.99}SnTeSe$-$Na_{0.01}$ HEA. This band structure modification and increase in DOS significantly enhanced the absolute Seebeck coefficient from 159.67 µV/K to 230 µV/K. This results in a virtuous improvement of 178% in the thermoelectric figure of merit, from 0.47 to 0.84 (Raphel et al. 2022).

6.7.4 Estimation of thermodynamic variables of high entropy alloys via first principle approach

The classical thermodynamic theories are formulated based on the assumptions that atoms in a solid solution are equal-sized hard spheres. However, these atoms are of different sizes depending upon the elemental concentration. Yet atoms with different sizes can create certain uncertainty or randomness in the lattice that can increase the configurational entropy of the system. The effect of such enhancement in the configurational entropy is less significant in conventional alloys when the concentration of the solute atoms is small compared to the parent materials. In multi-component alloys such as HEA, however, all the constituent elements are principal elements. The randomness due to atomic size differences among the constituent elements will be much higher for such alloys, resulting in a notable enhancement in the configurational entropy. These uncertainties or randomness in the lattice increases as the atomic size difference and the atomic concentration increase (Miracle and Senkov 2017). Mansoori et al. proposed an equation describing the effect of atomic size difference on excess entropy in a solid solution. Also, attempts were made to estimate excess entropy resulting from

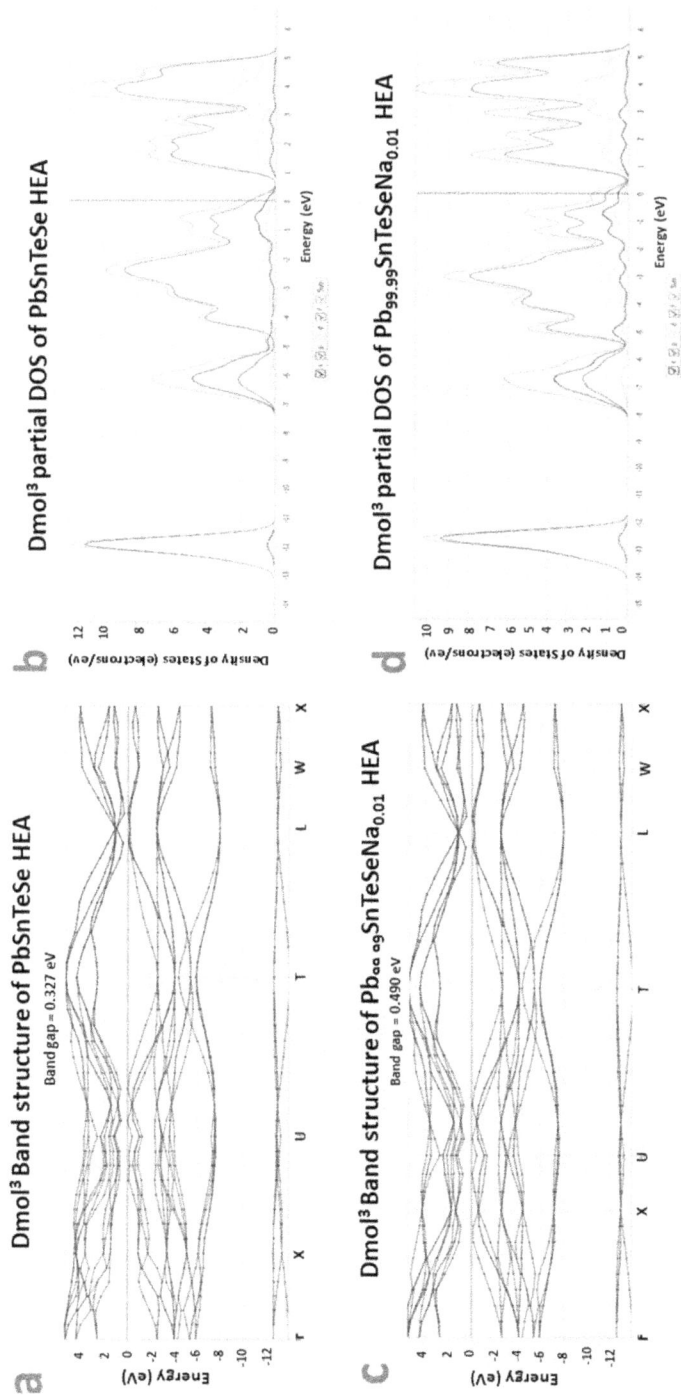

Figure 6.5 (a) Electronic band structure of PbSnTeSe (b) Electron density of states of PbSnTeSe (c) Electronic band structure of $Pb_{99.99}SnTeSeNa_{0.01}$ (d) Electron density of states of $Pb_{99.99}SnTeSeNa_{0.01}$.

atomic size differences on ternary amorphous alloys (Takeuchi and Inoue 2000). Thus, it can be inferred that the configuration entropy can be notably increased in multi-component alloys like HEA, and the actual entropy can be higher than the theoretical ideal entropy estimation. The further addition of alloying elements or dopants can act as atomic point defects and could significantly affect configurational entropy and alloy stability.

Attempts were made to understand the effect of such thermodynamic parameters such as enthalpy of mixing (ΔH), the entropy of mixing (ΔS), Gibbs free energy (ΔG) of the alloy system was carried out using DFT first principle approach via DMol^3 software module. Thermodynamic variables of the synthesized PbSnTeSe and $PbSn_{0.875}TeSeBi_{0.125}$ HEA, such as enthalpy (ΔH), entropy (ΔS), free energy (ΔG) as a function of temperature (T), were estimated using the DMol^3 software module (Raphel et al. 2021b). Figure 6.6 portrays the effect of Bismuth (Bi) doping in thermodynamic properties of PbSnTeSe HEA. The results indicate a notable change in the mixing entropy of the alloy over the entire measurement range. This can be attributed to the increased microscopic configurations resulting from the

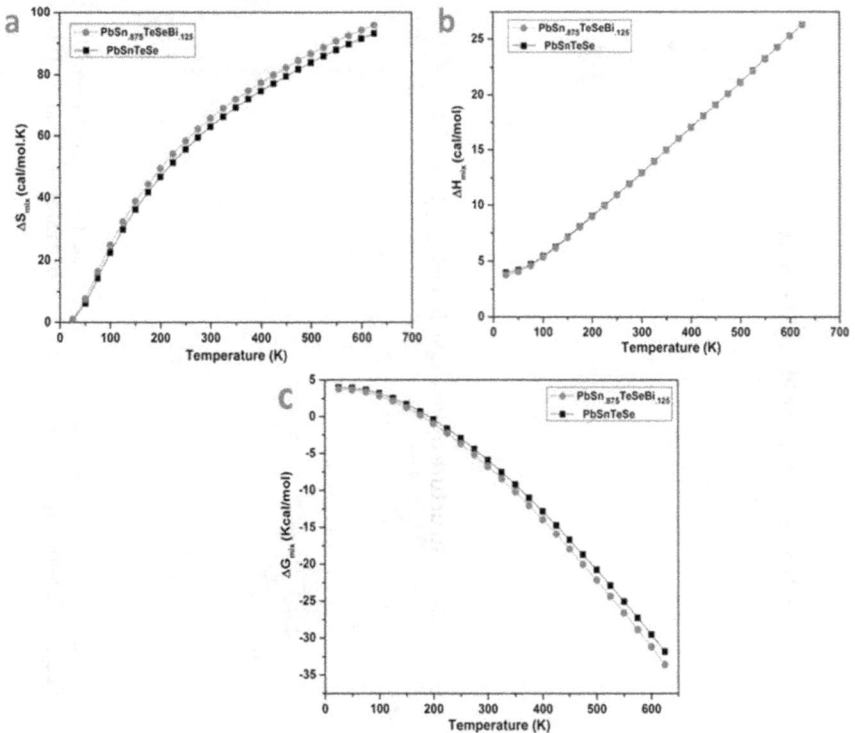

Figure 6.6 Temperature dependence of (a) mixing entropy (b) mixing enthalpy, and (c) Gibbs free energy of PbSnTeSe and $PbSn_{0.875}TeSeBi_{0.125}$ HEA.

doping of the Bi atom in the quaternary random solid solution consisting of Pb, Sn, Te, and Se atoms. Another interesting observation is that mixing enthalpy of the base and doped alloy exhibits similar temperature dependence and near-equal value over the entire measuring range. Thus, it decreases Gibbs free energy and will thereby improve alloy stability. In Bi-doped PbSnTeSe HEA, there is a 2.71 kcal/mol (4.3%) increase in the configurational entropy and a decrease of 0.8476 cal/mol-K (14.31%) in the Gibbs free energy of the system. This entropy engineering via Ag doping significantly achieves a highly desirable ZT of 0.71 and retains a stable mono-phase in Bi-doped PbSnTeSe HEA.

6.8 CONCLUSION

The concept of entropy engineering via multi-component alloy design showcases a novel path for the development of tailored material chemistry for functional and structural applications. Developing such alloys with higher ZT can be considered the future for thermoelectric research. Among the various synthesis routes for thermoelectric high entropy alloy development, the powder metallurgical route, which combines mechanical alloying and spark plasma sintering, is emerging as a swift route for single-phase alloy development with better homogeneity. Further, the potential doping of such alloys can further fine-tune the thermoelectric performance via band engineering and increase the electron density of states. The Na doping in PbSnTeSe HEA modifies the band structure by remarkably increasing the bandgap and finetuning the DOS, thereby enhancing the absolute Seebeck coefficient and the thermoelectric figure of merit. The doping of Bi in PbSnTeSe HEA produced a significant enhancement in configurational entropy, improving the alloy's stability. The density function theory-based first principle studies can provide more insights into the crystal structure, band structure, thermodynamic parameters, and alloy stability.

REFERENCES

Delley, B. (1990) B An all-electron numerical method for solving the local density functional for polyatomic molecules. *J. Chem. Phys.*, 92, 508–17.

Delley, B. (2000) B From molecules to solids with the $DMol^3$ approach. *J. Chem. Phys.*, 113, 7756–64.

Dong, W., Z. Zhou, L. Zhang, M. Zhang, P.K. Liaw, G. Li, and R. Liu (2018) Effects of Y, Gd, Cu, and Al addition on the thermoelectric behavior of CoCrFeNi high entropy alloys. *Metals*, [online] 8(10), 781. https://doi.org/10.3390/met8100781.

Fan, Z., H. Wang, Y. Wu, X. J. Liu, and Z. P. Lu (2016) Thermoelectric high-entropy alloys with low lattice thermal conductivity. *RSC Adv.*, 6, 52164.

Fan, Z., H. Wang, Y. Wu, X. J. Liu, and Z. P. Lu (2017) Thermoelectric performance of PbSnTeSe high entropy alloys. *Mater. Res. Lett.*, 5(3), 187–194.

Gayner, C., K. K. Kar, and W. Kim (2018) Recent progress and futuristic development of PbSe thermoelectric materials and devices. *Mater. Today Energy*, 9, 359–76.

Han, K., H. Jiang, T. Huang, and M. Wei (2020) Thermoelectric properties of CoCrFeNiNb$_x$ eutectic high entropy alloys. *Crystals*, 10(9), 762.

Heremans, J. P., V. Jovovic, E. S. Toberer, A. Saramat, K. Kurosaki, and A. Charoenphakdee (2008) Enhancement of thermoelectric efficiency in PbTe by distortion of the electronic density of states. *Science*, 321, 554–7.

Heyd, J., G. E. Scuseria, and M. Ernzerhof (2003) Hybrid functionals based on a screened Coulomb potential. *J. Chem. Phys.*, 118, 8207–15.

Jeffrey Snyder, G., and Eric S. Toberer (2008) Complex thermoelectric materials. *Nat. Mater.*, 7, 105–14.

Meng, H., M. An, T. Luo, and N. Yang (2020) Thermoelectric applications of chalcogenides. In Xinyu Liu, Sanghoon Lee, Jacek K. Furdyna, Tengfei Luo, Yong Zhang (Eds.), *Chalcogenide: From 3D to 2D and Beyond*, Woodhead Publishing.

Miracle, D. B., and O. N. Senkov (2017) A critical review of high entropy alloys and related concepts. *Acta Mater.*, 122, 448–511.

Perdew, J. P., K. Burke, and M. Ernzerhof (1996) Generalized gradient approximation made simple. *Phys. Rev. Lett.*, 77, 3865–3868.

Perumal, S., S. Roychowdhury, and K. Biswas (2016) High performance thermoelectric materials and devices based on GeTe. *J. Mater. Chem. C*, 4, 7520–7536.

Qin, Y., Y. Xiao, D. Wang, B. Qin, Z. Huang, and L. D. Zhao (2019) An approach of enhancing thermoelectric performance for p-type PbS: Decreasing electronic thermal conductivity. *J. Alloys Compd.*, 820, 153453.

Raphel, A., A. K. Singh, P. Vivekanandhan, and S. Kumaran (2021a) Thermoelectric performance of nanostructured PbSnTeSe high entropy thermoelectric alloy synthesized via spark plasma sintering. *Physica B: Condensed Matter.*, 622, 413319.

Raphel, A., P. Vivekanandhan, and S. Kumaran (2020) High entropy phenomena induced low thermal conductivity in BiSbTe$_{1.5}$Se$_{1.5}$ thermoelectric alloy through mechanical alloying and spark plasma sintering. *Mater. Lett.*, 269, 127672.

Raphel, A., P. Vivekanandhan, A. K. Singh, and S. Kumaran (2021b) High entropy stabilization and band engineering driven high figure of merit in nanostructured PbSn$_{0.875}$TeSeBi$_{0.125}$ alloy. *J. Solid State Chem.*, 303, 122531.

Raphel, A., P. Vivekanandhan, A. K. Singh, and S. Kumaran (2022) Tuning figure of merit in Na doped nanocrystalline PbSnTeSe high entropy alloy via band engineering. *Mater. Sci. Semicond. Process.*, 138, 106270.

Shafeie, S., S. Guo, Q. Hu, H. Fahlquist, P. Erhart, and A. Palmqvist (2015) High-entropy alloys as high-temperature thermoelectric materials. *J. Appl. Phys.*, 118, 184905.

Shen, Q., L. Chen, T. Goto, T. Hirai, J. Yang, G. P. Meisner, and C. Uher (2001) Effects of partial substitution of Ni by Pd on the thermoelectric properties of ZrNiSn-based half Heusler compounds. *Appl. Phys. Lett.*, 79(25), 4165–4171.

Takeuchi, A., and A. Inoue (2000) Calculations of mixing enthalpy and mismatch entropy for ternary amorphous alloys. *Materials Transactions, JIM*, [online] 41(11), 1372–1378. https://doi.org/10.2320/matertrans1989.41.1372.

UN Environment (2021) *Renewables 2021 Global Status Report*. UNEP - UN Environment Programme. http://www.unep.org/resources/report/renewables-2021-global-status-report

Wang, Z., D. Wang, Y. Qiu, J. He, and L. D. Zhao (2019) Realizing high thermoelectric performance of polycrystalline SnS through optimizing carrier concentration and modifying band structure. *J. Alloys Compd.*, 789, 485–492.

Zhang, X., and L.-D. Zhao (2015) Thermoelectric materials: Energy conversion between heat and electricity. *J. Materiomics*, 1, 92–105.

Zhao, L., Wang, J., Li, J., Liu, J., Wang, C., and Wang, J. (2019) High thermoelectric performance of Ag-doped SnTe polycrystalline bulks via the synergistic manipulation of electrical and thermal transport. *Phys. Chem. Chem. Phys.*, 21, 17978–84.

Zhao, L.-D., C. Chang, G. Tan, and M. G. Kanatzidis. (2016) SnSe: A remarkable new thermoelectric material. *Energ. Environ. Sci.*, 9, 3044–3060.

Chapter 7

Augmented and virtual reality incorporation in the manufacturing industry 4.0

M. Kiruthiga Devi, N. Kanya, N. Ethiraj and S. Sendilvelan

M.G.R Educational and Research Institute, Chennai, Tamil Nadu, India

CONTENTS

7.1 INTRODUCTION

Virtual reality (VR) mimics a product or environment digitally in the context of manufacturing and product design, frequently allowing the user to interact and become fully immersed. In contrast to VR, which simulates a virtual environment, AR projects a digital object or piece of information onto a real-world background [1]. To provide more information, AR places a digital layer on top of the physical world. This has potential applications across specialties and the capacity to offer complicated technical training, making it useful in fields like mechanical engineering [2]. AR systems share many of the same components as VR systems. AR systems, on the other hand, are not designed to immerse users in these virtual environments. With the use of AR programs, reality can be enhanced on a computer, a smartphone, or a tablet. This can be done for all of the senses, but it mostly focuses on the visual sense by using computer-generated content to add images and films to real items. The area of engineering that deals with the conception, creation, and maintenance of machines are known as mechanical engineering. One of the branches that rely largely on manual labour to complete the work is this one. However, Industry 4.0 recommends automation as a fix for any sector that is heavily dependent on labour. The use of AR and VR in mechanical engineering can be advantageous [3]. Nevertheless, some procedures are overly intricate. For those procedures, AR gives engineers the option of taking instructions and making them always visible in their field of view [4].

Additionally, it enables the engineer to operate using a voice-activated system, freeing up their hands. AR glasses used to work according to the instructions are visually demonstrated using 3D working models, and the complete process, including outside assembly, may also be demonstrated. As the engineer does not have to walk around to verify the instructions or bring stuff in between the procedure, this can save a lot of time. The industries have tested this technique, and it has been successful for them [5].

7.2 EVOLUTION OF AR AND VR

- 1838: The stereoscope was created so that drawing something from two different angles and then viewing images with a separate eye can produce a 3D effect. Such an opportunity was given to the viewer by the stereoscope. The invention of this device paved the way for modern technologies such as cinematography and photography.
- 1949: Improvements to the stereoscope lead to the invention of the "lenticular stereoscope," the first portable 3D viewer and physical optics.
- 1901: For the first time, an AR-like technology is mentioned in a book by L. Frank Baum. The protagonist of the book, *The Master Key: An Electrical Fairy Tale*, is a young man who is fascinated by technology, particularly electricity [6].
- 1929: The "Link Trainer" flight simulator is created by Ed Link. Using pumps, valves, and other devices, this simulator gave the pilots a precise simulation of what it's like to fly an airplane. This was an effective attempt to use a VR prototype [7].
- 1935: In his book *Pygmalion Spectacles*, Stanley G. Weinbaum mentions a set of glasses that allow the wearer to experience virtual environments through holographic pictures, smell, touch, and taste. Science fiction has always foreshadowed what we have now.
- 1952: The first VR-like device with immersive multimodal technology is created by Morton Heilig Called Sensorama, it had a stereoscopic colour display, odour emitters, a sound system, and fans. Because the Sensorama can attract people into a wide-angle stereoscopic vision, Morton was able to capture the attention of the entire audience.
- 1968: Ivan Sutherland and his student Bob Sproull made "The Sword of Damocles," which is a VR head-mounted display. Ivan was already renowned for his work in making computer graphics, which helped him make this device. The viewpoint of the image was determined by the head-tracking data, and it showed wireframe rooms that had been created digitally.
- 1982: The Video place lab for VR is created by Myron Krueger. His research with virtual environments led to the development of his notion. To converse in "artificial reality" without using devices like

motion-tracking gloves, Myron founded the lab. All the hardware pieces required to put the user in the simulated environment were available at the Video place.

- 1990: The phrase "augmented reality" is coined by Tom Caudell. He developed a replacement for the diagrams that were used to direct field personnel while he was employed by Boeing. He suggested giving the staff head-mounted devices that projected the plane's blueprints onto wipeable boards. A computer system might be used to simply alter the presented images. CAVE-style VR devices would dominate the market for VR in the 1990s and establish a new standard. CAVEs are spaces where wall-mounted projectors are used to create immersive virtual environments. Users did not want to experience the negative of headaches and motion sickness from CAVEs.
- 1992: Movie premiere for *The Lawnmower Man*. The plotline was based on a fictional account of a scientist who treated a patient with mental illness using VR. This offered another illustration of how VR was naturally integrating into the media entertainment sector. Virtual Fixtures are developed by Louis Rosenberg. This AR system was created to enhance users' ability to carry out manipulating tasks. The U.S. Air Force Research Laboratory created the system, which was a traditional AR overlay on the user's information perception. A Fitts Law performance test was completed by the system, demonstrating how such systems can greatly boost human performance in specific jobs.
- 1995: Nintendo displays the VR-32, a proprietary product that will eventually become the Virtual Boy. At the Consumer Electronics Show, Nintendo proclaimed that their new product would provide users with a breathtaking virtual reality interaction. The firm took a big risk when it released the first home VR product, Virtual Boy, which became the first home VR gadget.
- 2000: Release of the ARToolKit. This ground-breaking computer-tracking library made it possible to develop AR applications. It is currently hosted on GitHub and is Open Source. EyeTap is a manufactured wearable. EyeTap adds computer-generated data by interpreting the user's eye as both a camera and a monitor. This year also saw the release of the first smartphone-augmented reality game, ARQuaqe. It needed a linked plastic gun, a laptop, and a head-mounted display, which the user would carry in their backpack, a head-movement tracker, and a GPS. This enormous structure weighed 16 kg, meaning that it was clear that it was not made for easy gameplay.
- 2012: Palmer Luckey unveils the initial version of the Oculus VR HMD. This was a massive apparatus that could project a 90-degree field of view 2D images and was so hefty that even required a counterbalance in the backside was revealed by him. Luckey made the decision that he needed to improve his equipment, making it lighter and more potent. Google also releases information about its head-mounted

optical display, Glass. Despite some unfavorable reviews, the device was a tremendous advancement for both the IoT sector and VR development. Early users had high expectations that were not met; and therefore Google had a lot to improve and develop [8].

- 2014: Morpheus, a project from Sony that will lead ultimately to PlayStation VR, is announced. This was a project to make VR glasses for the PlayStation 4. The engineers played around with their hardware, portable video players, and PlayStation Move controllers to make a prototype that worked perfectly. At the Game Developers Conference, they showed it off with a few games.
- 2017: Google and Apple have released their frameworks for AR. The most cutting-edge AR development tools are made available to developers by ARCore and ARKit so they may create lifelike AR experiences. In Figure 7.1, we explain the highlights of the Evolution of AR and VR

One of the most significant shifts can be noticed in the working environment, particularly in situations when people and robots collaborate on projects and where there is an overlap between the actual and virtual worlds [9]. Virtual animations and simulations of goods and processes, in addition to visualizations of raw and computed data, are among the most essential components of a Cyber-Physical System (CPS) [10]. When combined with the appropriate data, the applications of augmented reality (AR) and visual computing can be used to facilitate improved vertical integration in the factory [11]. AR is a novel way to assist employees and a major link between people and the Industry 4.0 environment as a way for people and machines to collaborate [12, 13]. AR is also a crucial link between people and the Industry 2.0 environment. AR has the potential to transform present-day workers into intelligent workers and operators of the future, provided that these workers have access to the appropriate technology. These intelligent workers are quickly becoming one of the most essential components of the Industry 4.0 environment [14, 15] due to their ability to make strategic decisions and flexibly address problems. Because it can display the appropriate data at the appropriate time, augmented reality (AR) serves as an essential link between the web database of the plant and the intelligent workers there [9, 16]. The difficulty lies in developing AR applications that may reduce production times and costs while also improving product quality [17]. Augmented reality can be used in conjunction with brain–computer interface (BCI), technology to control various activities. Through the use of BCI technology, direct communication between the brain and an external device is made possible. It is possible to manage the process and have an effect on various items in space based on the signals that are generated. One example of this would involve the movement of a mobile robot [18].

AR applications have the potential to help solve problems like workers not having access to enough information, workers not getting enough training,

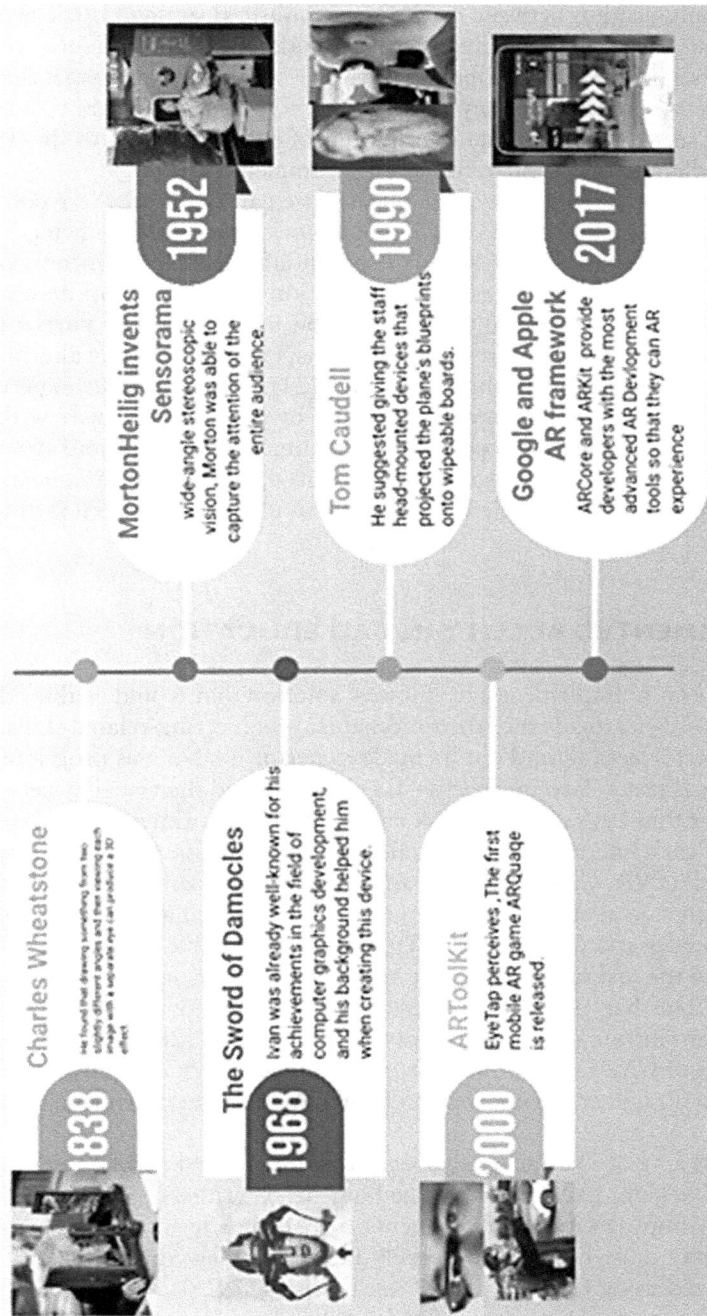

Figure 7.1 Evolution of AR and VR.

the gap between the actual problem and the solution that was planned, and poor communication between actors in an industrial setting [19]. The addition of real-time virtual information to the working environment of a worker has the potential to not only make their tasks simpler but also assist them in making better judgments. They will continue to play a significant role in the industrial sector for a considerable amount of time, regardless of the degree to which digitalization and automation are implemented.

Because the majority of worker injuries are caused by workers not having sufficient training, not having enough work experience, or being bored or distracted at work [20], AR has the potential to play a significant role in making industrial workplaces safer. In addition to the phases of design and production, augmented reality may also be useful in other parts of the product life cycle. To be more specific, it assists in marketing by altering the way in which clients feel about pre-sales [21]. The after-sales experience can also be improved by augmented reality by providing the user with AR manuals, instructions for repairs, or communication with after-sales sales personnel, among other uses. AR is also used extensively in educational contexts [22–24], which helps train people to use it in industrial settings in the future [9, 25].

7.3 AUGMENTED REALITY IN CAD EDUCATION

It was crucial to implement a technology solution that would enable educators and students to advance their mechanical engineering-related skills. The various performers should not be made to perform laborious programming or graphical tasks. Two interfaces—the program and the visual target—will be used for this. Let's say, if you ask us, that you are working on the development of a car's interior or an internal combustion engine. Although it won't be as deep as VR, CAD data with augmented reality will let you sit within your design as you are making it, giving you a more enhanced experience in viewing angles and product specs. Designers will be able to see through the design into the real surroundings by superimposing the digital data of design elements. This has essentially ushered with AR by designers and engineers to visualize various sets of prototypes and hypotheses [26]. A model for the utilization of CAD and AR in CAD Education is shown in Figure 7.2.

Steps to do numeric work follows from CAD to AR experiment:

- Modeling & Designing: Implementing augmented reality (AR) can be challenging, particularly in medium-sized businesses, if you wish to superimpose virtual 3D elements of reality. The models we want to display must be created digitally in three dimensions. The CADs are created using Catia as part of our project [27, 28].
- The scenario: A scenario can be defined using the Catia Composer program, where each step specifies the components that will be displayed and the information or instructional texts that will be provided

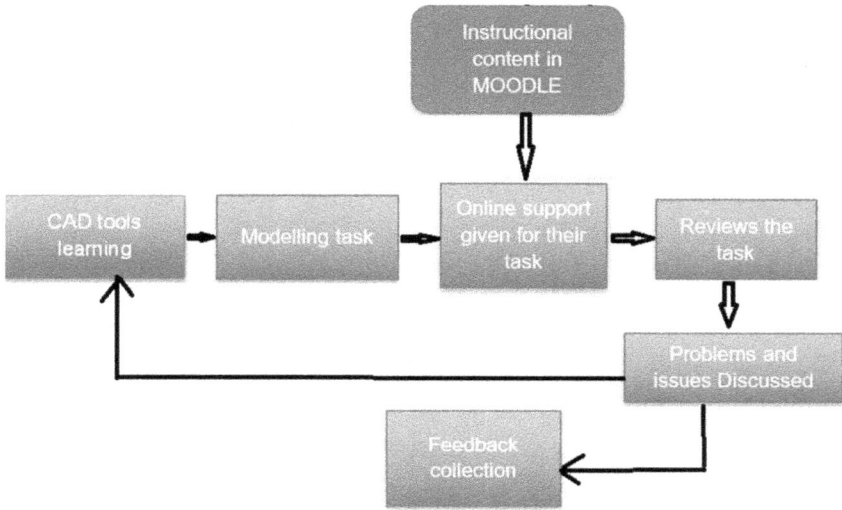

Figure 7.2 CAD and augmented reality in teaching.

to the user. This phase is crucial since the scenario needs to be progressive and accommodate the kids' abilities.

- The tracking model and export are defined as follows: You can specify the tracking model for each view in Diota for Composer. Different project export formats are available with this application, including. Diota player for usage on Surface tablets and Diota project for use on Hololens glasses [29]. The tracking model is transformed into a mesh in both scenarios, and we must ascertain its fineness.
- Utilization of AR: Using the Diota Player, the scenario is played after the export has been loaded and completed into the final user interface (Figure 7.3). The area of application is mentioned in Figure 7.4. The equipping methods are shown in Figure 7.5.

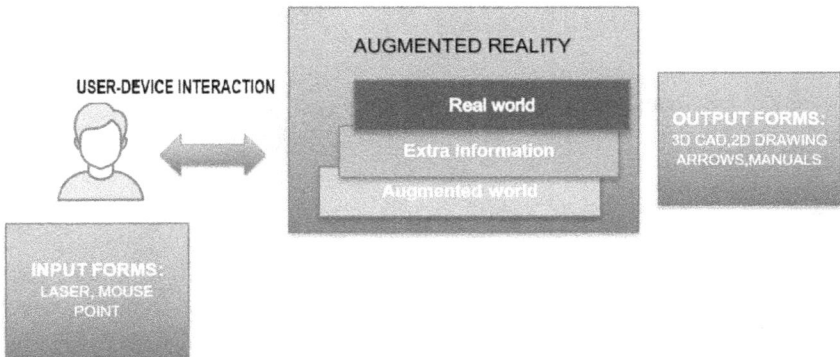

Figure 7.3 Workflow from CAD to AR.

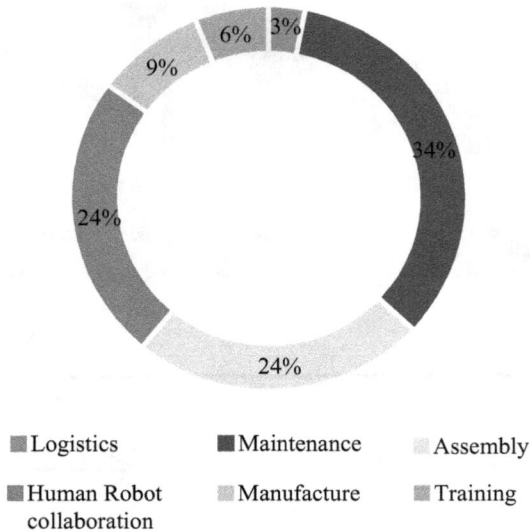

Figure 7.4 **Area of application.**

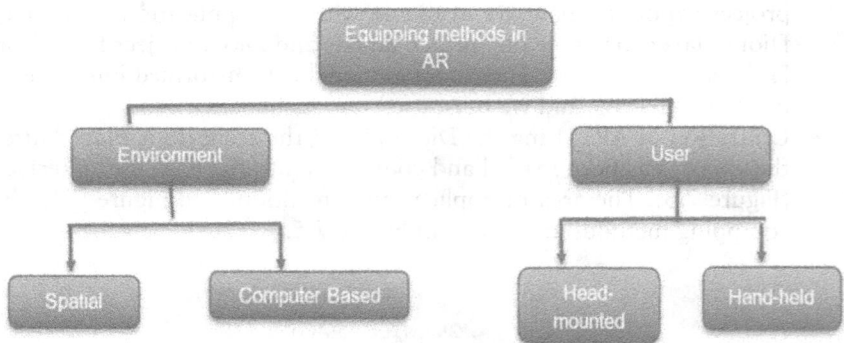

Figure 7.5 **Equipping methods in AR.**

7.4 EFFECTS OF TRACKING

In tracking, a real element must be linked by identification to a predefined virtual element so that the whole virtual overlay can be found in space. Every view can have its tracking model, which will be used as a guide throughout the whole augmented representation. To use a software solution that does not require programming and works with the Dassault System CAD solution so that instructors and students can use it without having to learn a lot about programming, spend a lot of time programming, or lose touch with their main tools. Only the Diota solution among the various platforms permits this. It provides tracking by identifying the shape of 3D parts. The Hololens glasses' numerous cameras or a Surface tablet's primary camera can both be used by the Diota Player to locate the actual tracking part(s) specified in the Diota for Composer software. This has various benefits over 2D targets, including the following: The tracking functions no matter where the user is located in the actual system. In comparison to genuine parts, virtual parts may be positioned in 3D exactly. By identifying the shape of 3D parts, it provides tracking [30]. The Diota Player recognizes the actual tracking part(s) specified in Diota for Composer, whether through the numerous cameras of Hololens glasses or through the primary camera of a Surface tablet. This has various advantages over 2D targets, including the fact that the tracking is effective regardless of where the user is concerning the actual system [31]. Also, unlike real parts, virtual parts can be placed more accurately in 3D.

For this process to work, a very accurate 3D model of the parts is required. However, there is a paradox here: if the tracking model is too simple, it can't be recognized uniquely (for example, a part of the tube is not unique in the real model), but if it's too complex, it will be very heavy, which will slow down the scenario, and make the parts more diffraction-limited. We tried scanning the gearbox housing to make a cloud of 3D points to get around the problem of complicated modeling or not having a digital representation of the part [32]. When used as a tracking model, as planned, this scan works perfectly. Since we do not need to show the casing in the scenario about how a gearbox works or how to take it apart or put it back together, the part does not look good, but this is not a problem.

- **Problem with tracking aspects on a surface table:** The tablet's Diota Player only takes into account the tracking model as a reference. This dictates that the full model must always be in the tablet's camera's range of vision [33]. This restriction may present difficulties, particularly when we have to take a step back or move closer to see the details. Depending on the scenario views, different reference parts (tracking models) might be defined as a partial solution to this issue. When we wish to portray the machine tool as a whole, the machine's case is employed initially. Then, in a subsequent phase, it was necessary to move closer to observe the interior of the machine like its axes, internal components, etc., for which another tracking object called an internal plate was utilized as shown in Figure 7.6.
- Glasses constantly scan the entire area (the room, tables, seats, etc.). This enables excellent virtual representation stability after hooking, regardless of whether the tracking model is still discernible by the

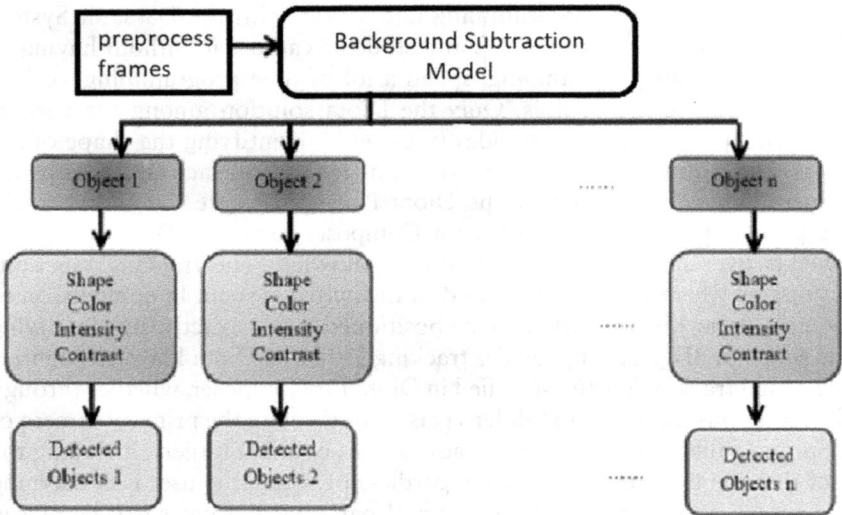

Figure 7.6 Multiple object tracking.

glasses' cameras. However, because a system's location would fluctuate about its environment, this technique does not apply to it [34, 35]. As a result, a tracking calibration may no longer be valid if the design of the room where the system under study is located changes as shown in Figure 7.7.

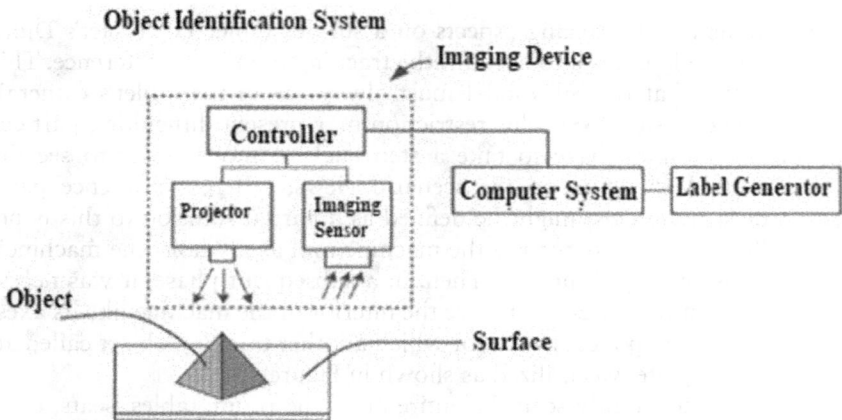

Figure 7.7 Identification of the object with label.

7.5 CONCLUSION

The initial contribution of augmented reality to help higher engineering education is to acquaint tomorrow's engineers with a new emerging technology they will probably encounter in their future careers, allowing them to see its potential and envision its applications in the industry of the future. While the students' interest in augmented reality technologies was evident each time, it's crucial to make a distinction between their inherent capacity for understanding complicated systems and their specific comprehension of how a given system works. Even if augmented reality makes it easier to share information about a system, this does not change how well it can think critically right away. It makes more sense to see this technology as a promising support tool that will help lead the student, make it easier to analyze a mechanism, and be more effective, rather than as a magic tool that will make the learner smarter. But right now, it doesn't seem likely that AR will completely replace all traditional tools.

REFERENCES

[1] Radianti, Jaziar; Majchrzak, Tim A.; Fromm, Jennifer; Wohlgenannt, Isabell. A systematic review of immersive virtual reality applications for higher education: Design elements, lessons learned, and research agenda. *Comput. Educ.* 2020, 147, 103778.

[2] Chavez, B.; Bayona, S. Virtual Reality in the Learning Process. Trends and Advances in Information Systems and Technologies. In *Presents the outcomes of the 2018 World Conference on Information Systems and Technologies (WorldCIST'18) held in Naples, Italy, from March 27 to 29, 2018.* Springer International Publishing, 2018, pp. 1345–1356.

[3] Chen, Y.-L. The effects of virtual reality learning environment on student cognitive and linguistic development. *Asia Pac. Educ. Res.* 2016, 25(4), 637–646.

[4] Fosnot, C.T.; Perry, R.S. Constructivism: A psychological theory of learning. In *Constructivism: Theory, Perspectives, and Practice* 1996, 2nd Edition, pp. 8–33. Copyright © 2005 by Teachers College, Columbia University. All rights reserved. ISBN 0-8077-4570-0.

[5] Garrett, L. Dewey, Dale, and Bruner: Educational philosophy, experiential learning, and library school cataloging instruction. *J. Educ. Libr. Inf. Sci.* 1997, 38(2), 129–136.

[6] Sato, M.; Peaucelle, D. Conservatism reduction for linear parameter-varying control design facing inexact scheduling parameters illustrated on flight tests. *Int. J. Robust Nonlinear Control.* 2020, 30, 6130–6148. https://doi.org/10.1002/rnc.5037

[7] Frank Baum, L. *The Master Key: An Electrical Fairy Tale.* Introduction by David L. Green and Douglas G. Greene. Westport, CT: Hyperion Press, 1974.

[8] Scaravetti, D.; François, R. Implementation of augmented reality in a mechanical engineering training context. *Computers, MDPI* 2021, 10(12), 163. https://doi.org/10.3390/computers10120163. ⟨hal-03514380⟩

[9] Reljić, V.; Milenković, I.; Dudić, S.; Šulc, J.; Bajči, B. Augmented reality applications in Industry 4.0 environment. *Appl. Sci.* 2021, 11, 5592. https://doi.org/10.3390/app11125592

[10] Lee, J.; Bagheri, B.; Kao, H.A. A cyber-physical systems architecture for Industry 4.0-based manufacturing systems. *Manuf. Lett.* 2015, 3, 18–23.

[11] Posada, J.; Toro, C.; Barandiaran, I.; Oyarzun, D.; Stricker, D.; De Amicis, R.; Pinto, E.B.; Eisert, P.; Döllner, J.; Vallarino, I. Visual Computing as a Key Enabling technology for Industrie 4.0 and industrial internet. *IEEE Comput. Graph. Appl.* 2015, 35, 26–40.

[12] Langfinger, M.; Schneider, M.; Stricker, D.; Schotten, H.D. Addressing security challenges in industrial augmented reality systems. In *Proceedings of the IEEE 15th International Conference on Industrial Informatics (INDIN)*, Emden, Germany, 24–26 July 2017; pp. 299–304.

[13] Belkadi, F.; Dhuieb, M.A.; Aguado, J.V.; Laroche, F.; Bernard, A.; Chinesta, F. Intelligent assistant system as a context-aware decision-making support for the workers of the future. *Comput. Ind. Eng.* 2020, 139, 105732.

[14] Kolberg, D.; Zühlke, D. Lean automation enabled by Industry 4.0 technologies. *IFAC-PapersOnLine* 2015, 28, 1870–1875.

[15] Longo, F.; Nicoletti, L.; Padovano, A. Smart operators in industry 4.0: A human-centered approach to enhance operators' capabilities and competencies within the new smart factory context. *Comput. Ind. Eng.* 2017, 113, 144–159.

[16] Lucke, D.; Constantinescu, C.; Westkämper, E. Smart factory—A step towards the next generation of manufacturing. In *Manufacturing Systems and Technologies for the New Frontier*; Mitsuishi, M., Ueda, K., Kimura, F., Eds.; London, UK: Springer, 2008; pp. 115–118.

[17] Cardoso, L.F.D.S.; Mariano, F.C.M.Q.; Zorzal, E.R. A survey of industrial augmented reality. *Comput. Ind. Eng.* 2020, 139, 106159.

[18] Chi, H.L.; Kang, S.C.; Wang, X. Research trends and opportunities of augmented reality applications in architecture, engineering, and construction. *Autom. Constr.* 2013, 33, 116–122.

[19] Paszkiel, S. *Analysis and Classification of EEG Signals for Brain Computer Interfaces*; Cham, Switzerland: Springer, 2020; ISBN 978-3-030-30580-2.

[20] Dušan T.; Bojan T. The application of augmented reality technologies for the improvement of occupational safety in an industrial environment. *Comput. Ind.* 2017, 85, 1–10.

[21] Fraga-Lamas, P.; Fernández-Caramés, T.M.; Blanco-Novoa, Ó.; Vilar-Montesinos, M.A. A review on industrial augmented reality systems for the Industry 4.0 shipyard. *IEEE Access* 2018, 6, 13358–13375.

[22] Bower, M.; Howe, C.; McCredie, N.; Robinson, A.; Grover, D. Augmented Reality in education—cases, places and potentials. *Educ. Media Int.* 2014, 51, 1–15.

[23] Cheng, K.H.; Tsai, C.C. Affordances of augmented reality in science learning: Suggestions for future research. *J. Sci. Educ. Technol.* 2013, 22, 449–462.

[24] Andújar, J.M.; Mejías, A.; Márquez, M.A. Augmented reality for the improvement of remote laboratories: An augmented remote laboratory. *IEEE Trans. Educ.* 2011, 54, 492–500.

[25] Fraga-Lamas, P.; Fernández-Caramés, T. M.; Blanco-Novoa, Ó.; Vilar-Montesinos, M. A review on industrial augmented reality systems for the Industry 4.0 shipyard. *IEEE Access* 2018, 6, 13358–13375. https://doi.org/10.1109/ACCESS.2018.2808326.

[26] Cometti, C.; Païzis, C.; Casteleira, A.; Pons, G.; Babault, N. Effects of mixed reality head-mounted glasses during 90 minutes of mental and manual tasks on cognitive and physiological functions. *PeerJ*. 2018, 6, e5847. https://doi.org/10.7717/peerj.5847. PMID: 30416883; PMCID: PMC6225835.

[27] Duh, H.B.L.; Klopfer, E. Augmented reality learning: New learning paradigm in co-space. *Comput. Educ.* 2013, 68, 534–535.

[28] Johnson, L.; Adams Becker, S.; Cummins, M.; Estrada, V.; Freeman, A.; Hall, C. NMC Horizon Report: 2016 Higher Education Edition; The New Media Consortium: Austin, TX, USA, 2016.

[29] Cohen, L.; Duboé, P.; Buvat, J.; Melton, D.; Khadikar, A.; Shah, H. Augmented and Virtual Reality in Operations; Capgemini Research Institute Report; Capgemini Research Institute, 2018.

[30] Cearley, D.; Burke, B. Top 10 Strategic Technology Trends for 2019. 2018. Available online: https://www.gartner.com

[31] Wang, M.; Callaghan, V.; Bernhardt, J.; White, K.; Penarios, A. Augmented reality in education and training: Pedagogical approaches and illustrative case studies. *J. Ambient. Intell. Humaniz. Comput.* 2018, 9, 1391–1402.

[32] Webel, S.; Bockholt, U.; Engelke, T.; Peveri, M.; Olbrich, M.; Preusche, C. Augmented reality training for assembly and maintenance skills. In *Proceedings of the BIO Web of Conferences*, EDP Sciences, Montpellier, France, 15–16 December 2011; Volume 1, p. 97.

[33] Webel, S.; Uli, B.; Engelke, T.; Gavish, N.; Olbrich, M.; Preusche, C. An augmented reality training platform for assembly and maintenance skills. *Robot. Auton. Syst.* 2013, 61, 398–403.

[34] Gavish, N.; Gutiérrez, T.; Webel, S.; Rodríguez, J.; Peveri, M.; Bockholt, U.; Tecchia, F. Evaluating virtual reality and augmented reality training for industrial maintenance and assembly tasks. *Interact. Learn. Environ.* 2015, 23, 778–798.

[35] Tümler, J.; Doil, F.; Mecke, R.; Paul, G.; Schenk, M.; Pfister, E.; Huckauf, A.; Bockelmann, I.; Roggentin, A. Mobile augmented reality in industrial applications: Approaches for solution of user-related issues. In *Proceedings of the 7th IEEE and ACM International Symposium on Mixed and Augmented Reality (ISMAR 08)*, Cambridge, UK, 15–18 September 2008; Volume 8, pp. 87–90.

Chapter 8

Modeling and optimization of process parameter for fatigue strength improvement by selective laser melting of AlSi$_{10}$Mg

Mudda Nirish and R. Rajendra

Osmania University, Hyderabad, Telangana, India

CONTENTS

8.1 INTRODUCTION

At present, modern industry also requires manufacturing geometrically complex shapes with reductions in cost, time, and weight [1, 2]. These results are mainly possible in metal additive manufacturing (AM) using powder with a laser power source as selective laser melting (SLM), also known as direct metal laser melting (DMLM), to build layer upon layer [3, 4]. SLM was constructed on printed specimens using process parameters such as scanning speed, laser power, and traversing every layer in the *x-y* plane [5, 6]. After each layer, the piston was lowered to allow the melted next layer of powder to be applied; this process was repeated several times until the finished part was obtained [7, 8]. When compared with other SLM-produced materials, AlSi$_{10}$Mg aluminium alloy powders have low density, high reflectivity, poor flowability, and high thermal conductivity [9, 10]. One of the most difficult challenges in the SLM fabrication of AlSi$_{10}$Mg alloy parts is minimizing porosity, and the majority of research has focused on how processing parameters affect porosity [11, 12]. Although SLM produces high-density components near the nominal density, as a result of oxides, gas bubbles and particles may become stuck due to process instabilities [13, 14]. The pores are unavoidable and can act as nuclei in the cracks, which can lead to reduced mechanical characteristics [15, 16]. The unmolten power particles, on the other hand, are indicated by irregular elongated pores, which are often caused by a lack of energy (such as hatch pattern defects) [17, 18]. The main impartment process parameters considered in the SLM printing process are layer thickness, scanning direction, build platform temperature, laser spot, scan speed, hatch distance, and laser power [19, 20].

In this chapter, thermal simulation of the part is presented and the part is affected by using different laser powers and scanning speeds [21, 22]. The process parameters for fatigue strength, hardness, and density without defects (porosity pores and cracks) were optimized, which can reduce the thermal warping of components made using the SLM AM technique [23, 24]. Some of the researchers have evaluated the main focus on mechanical properties (UTS, YS, and E%), microstructural characterization, and defects resulting from building orientation [25, 26].

8.2 EXPERIMENTAL PROCEDURE

Material. The $AlSi_{10}Mg$ alloy used for the SLM printing process of chemical composition is shown in Table 8.1. It was supplied by SLM solution group AG, Germany, and in the SLM printing process, the distribution of powder particle sizes ranges from 20 to 63 μm.

L-PBF of SLM Process. The $AlSi_{10}Mg$ alloy sample was printed by the SLM solution M280 2.0 L-PBF system (Germany). As shown in Figure 8.1, the SLM specification was a laser power of 400 Watts continuous Yb-fiber laser in an argon gas atmosphere and a build platform volume of 280 × 280 × 365 with various important key process parameters involved in the printing process.

The SLM input process parameters considered in Table 8.2 were laser power (two levels), scan speed (three levels), and hatch distance (three levels). Other parameters, such as build platform temperature of 1500C, laser spot diameter of 75 m, layer thickness of 30 m, scanning direction, and build orientation, were kept constant.

Table 8.1 Chemical composition of $AlSi_{10}Mg$ alloy

Al	Su	Fe	Cu	Mn	Mg	Zn	Ti	Ni	Pb	Sn	Other total
Balance	9.00– 11.00	0.55	0.05	0.45	0.20– 0.45	0.10	0.15	0.05	0.05	0.05	0.15

(a)

(b)

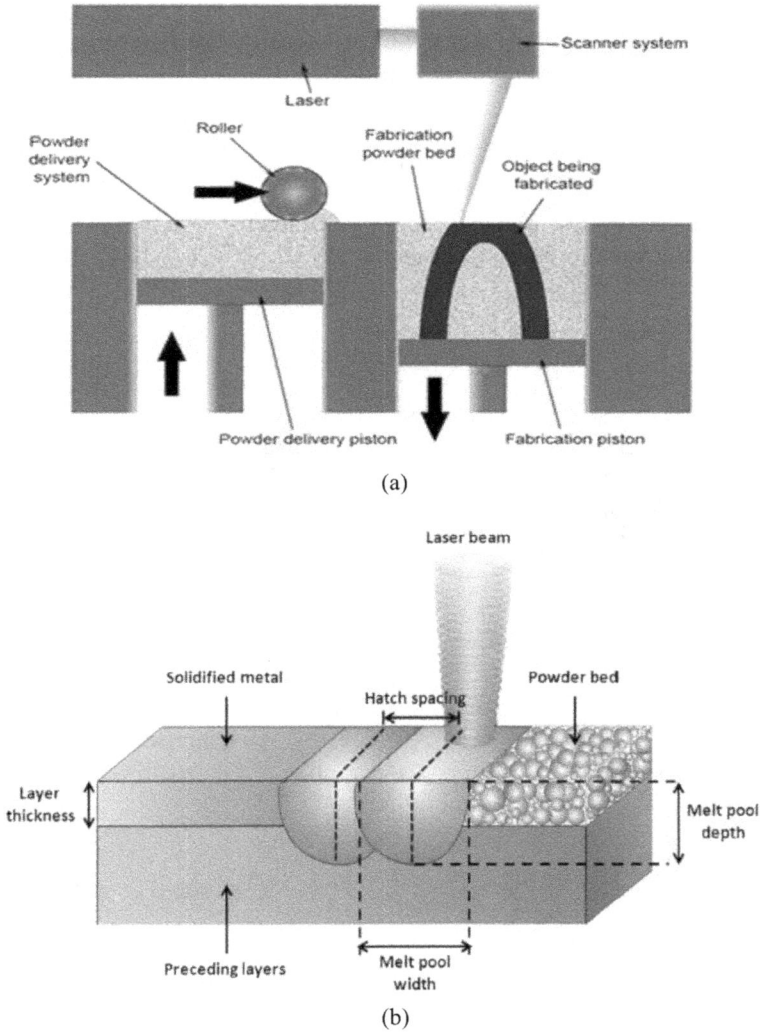

Figure 8.1 SLM printing process diagram and process parameter.

Table 8.2 Consider process parameter for SLM printing of AlSi$_{10}$Mg alloy

Laser power in Watts	Scan speed in mm/s	Hatch distance in μm	Specimen trails
250	400, 500 and 600	60, 80 and 100	T1, T2 and T3
300	400, 500 and 600	60, 80 and 100	T4, T5 and T6

SLM manufactured the part in accordance with ASTM standard E2948 for cyclic rotating bending (CRB) fatigue testing. The laser energy density was calculated by the following equation:

$$\frac{P}{v \times h \times t} \tag{8.1}$$

which is used for the above process parameters, which were calculated for T1 to T6 by the energy density as 347.2, 208.3, 138.8, 416.6, 250, 166.66 J/mm^3 as in Figure 8.2.

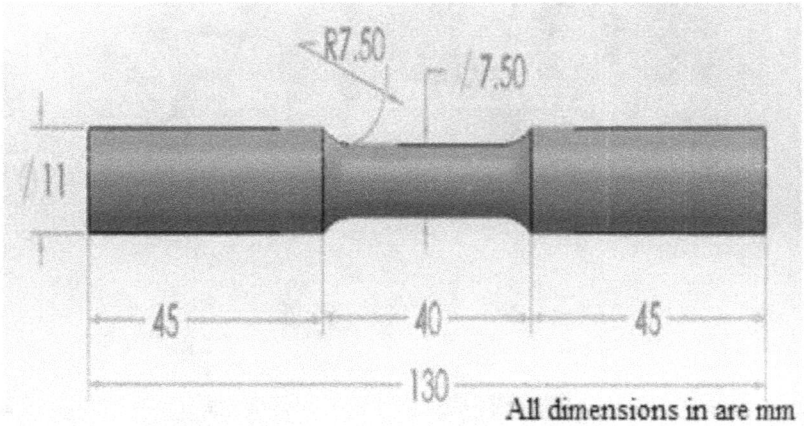

(a)

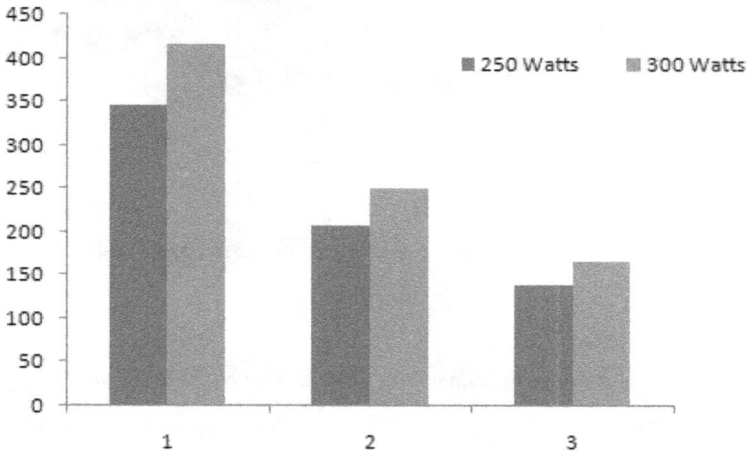

(b)

Figure 8.2 Fatigue test specimen its dimensions and energy density graph.

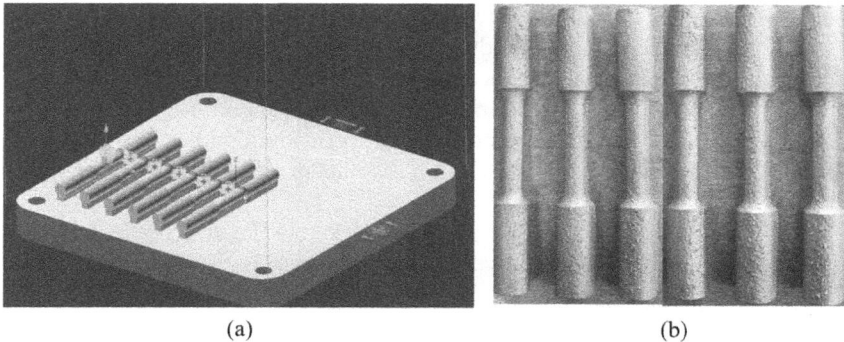

(a)   (b)

Figure 8.3 specimens on SLM build platform and after printed parts.

After giving the individual process parameters to specimens (such as; T1, T2, T3, T4, T5 and T6) in the SLM inbuilt software, we then generate the printing process to start the machine, as shown in Figure 8.3. The printing process is given two levels, i.e., consider the first level as 250 Watts and the second level as 300 Watts with the same scan speed range of (400, 500, and 600 mm/s) and hatching distance (60, 80, and 100 μm).

8.3 RESULTS AND DISCUSSION

In this work, we have determined the effect of porosity influence on fatigue behaviour with horizontal building orientation, which was conducted by a bending fatigue test, and also observed the microhardness and density parts of $AlSi_{10}Mg$ alloy produced by SLM. The experimental procedure is conducted by the following steps:

- Identify the SLM printing process key parameters.
- Develop the thermal simulation before the printing process.
- Consider the laser power in the upper 300 and the lower 250 Watts.
- Output performance is considered as fatigue strength, microhardness, and density.
- Develop the design of experiment for the optimal process parameter.
- Conduct the experiments as per the given process parameters on the fatigue test.

8.3.1 Thermal behaviour for optimal process parameter

Conducting a thermal analysis simulation prior to the SLM printing process saves material, money, and time. In that, thermal simulation stresses, such as displacement, temperature, plastic strain, Von Mises stress and other stresses, were developed. The thermal analysis simulations were conducted on individual specimens with different process parameters such as laser power in Watts (250 and 300), scan speed in mm/s (400, 500, and 600), and hatch distance in µm (60, 80, and 100), as mentioned above in Table 8.2. Figure 8.4 depicts the predicted maximum printing temperature in SLM of 246.6°C at T4 as well as the maximum displacement at T3 due to increased laser power with scan speed.

The highest Von Mises stress operated at 95.73 MPa at T5 (when increasing the scan speed) and the lowest plastic strain was developed at 0.33 at T3 (when increasing the hatch distance). During the optimization of this object, the purpose was to reduce the porosity and improve the strength. Finally, the simulation results obtained for optimal process parameters with less displacement and minimum operating temperature at T2 (laser power of 250 W, scan speed of 500 mm/s, and hatch distance of 80 µm) are shown in Figure 8.5.

8.3.2 Fatigue Performance

The fatigue test, as shown in Figure 8.6, is used and found to have horizontal building orientation specimen fatigue strength of SLM manufactured by $AlSi_{10}Mg$ alloy. The SLM-AM part building directions are the most important in terms of fatigue strength [10, 11]. From the fatigue failure data, it can be observed that when using high laser power with a low scan rate, the specimen's defects, pores, and porosity were observed after the test crack on part surfaces, and defects can have a harmful impact on fatigue strength.

The samples were tested by having a constant load applied to determine suitable process parameters based on the fatigue strength (such as fatigue life). The motor speed was calculated by the time the sample was broken and the number of cycles in time, as shown in Table 8.3. The fatigue strength calculation is

- The motor rpm is 2880,
  1 minute = 2880 rev (i.e., 1 second = $\dfrac{2880}{60}$ = 48 revolutions per second),
- The number of life cycles = time taken in second(s) × 48 revolutions per second and
- Length of the shaft (L) is 190 mm,
- Diameter of the shaft (d) is 15 mm,
- Weight (W) = 10 × 9.81 = 98.1 N (applied load for fatigue test is 10 kg),
- Bending moment [13]

$$
\begin{aligned}
(M) &= L \times W \\
&= 48069 \text{ N mm}
\end{aligned}
$$

(8.2)

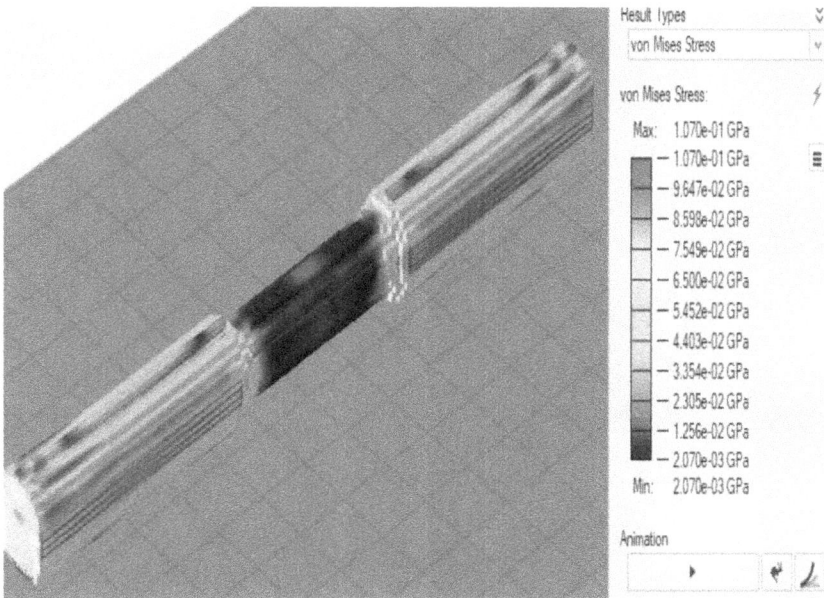

(a)

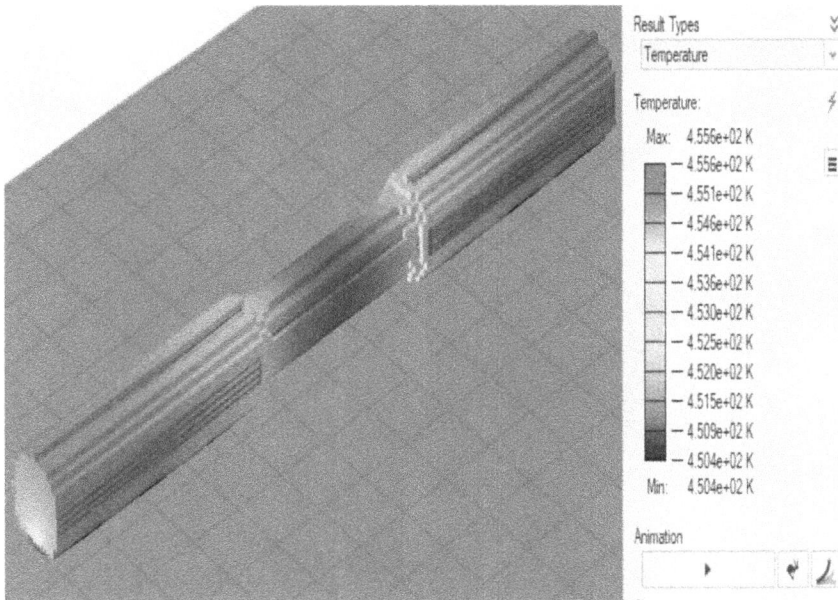

(b)

Figure 8.4 Thermal simulation as per given process parameter.

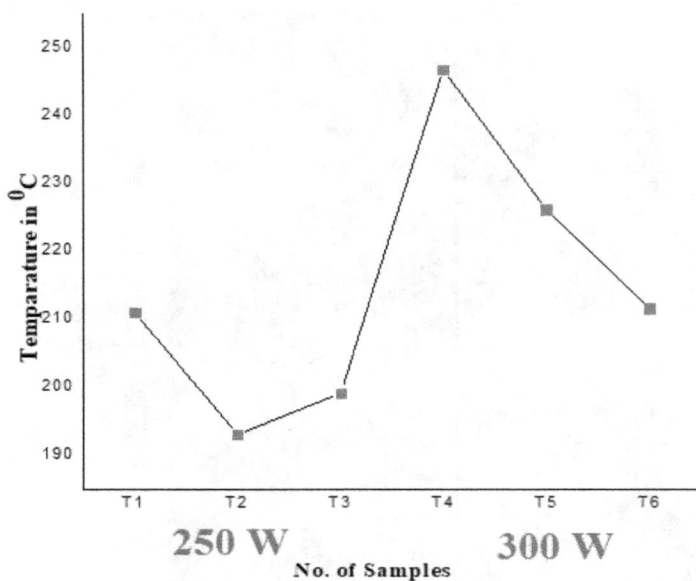

(a)

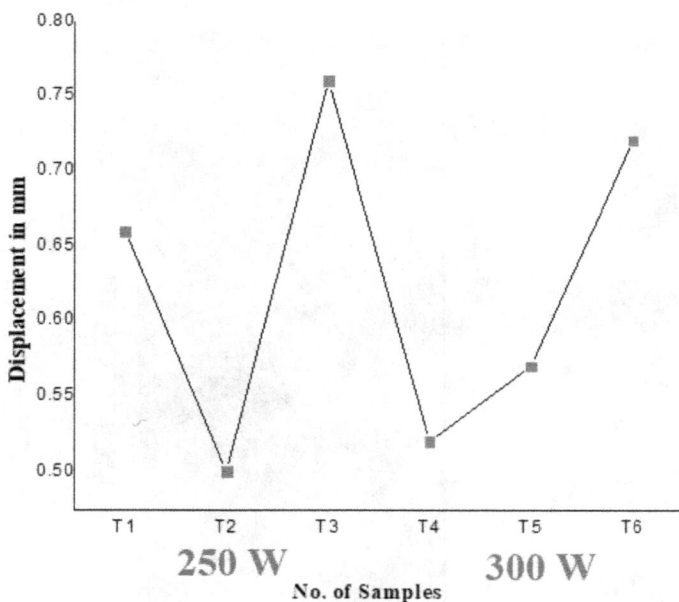

(b)

Figure 8.5 Thermal analysis simulation results and displacement values.

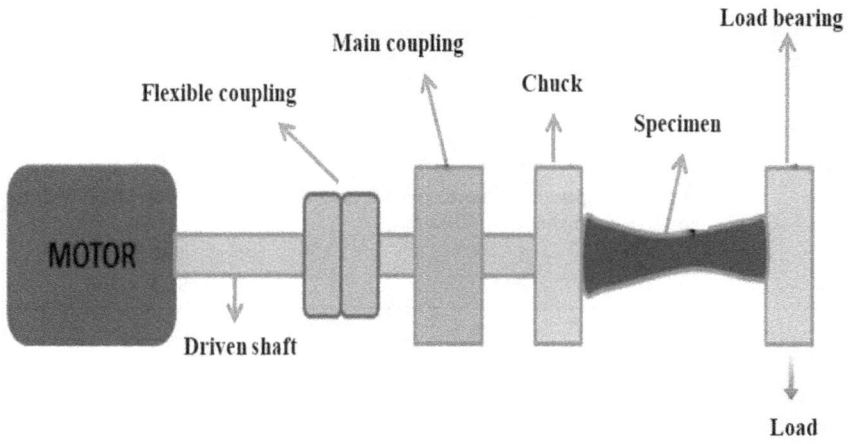

(a)

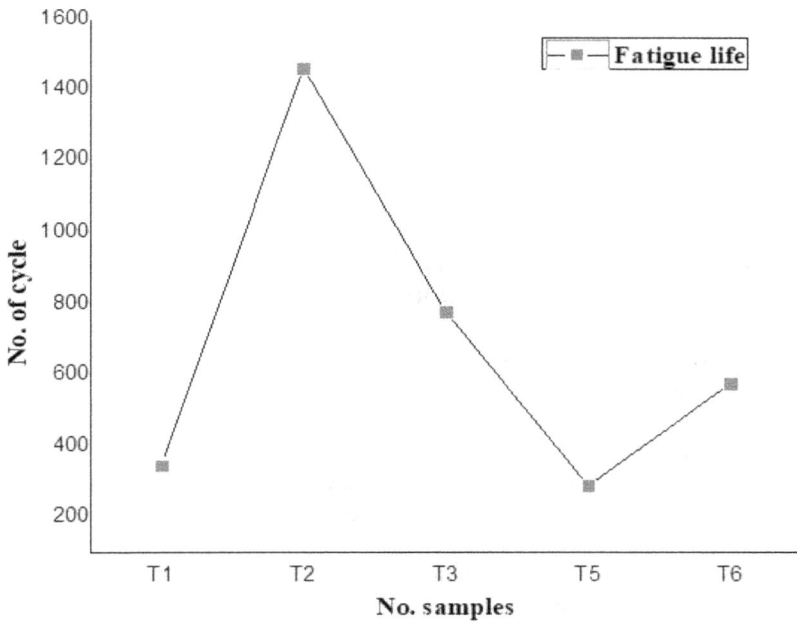

(b)

Figure 8.6 Fatigue testing and plotted vales at each specimen.

Table 8.3 After experimental test conducted of fatigue test results

Trail No.	Time taken in sec.	No. of cycles
T1	7.2	3.45×10^2
T2	30.6	1.46×10^3
T3	16.2	7.77×10^2
T4	Not printed specimen due to overlapping (high laser power of 300 Watts and low scan speed of 400 mm/s)	
T5	6	2.88×10^2
T6	12	5.76×10^2

and

- Bending stress [14]

$$\left(\sigma_b\right) = \frac{32M}{\pi d^3}$$
$$= 145.14 \text{MPa}$$

(8.3)

8.3.3 Microstructure, Hardness and Density

Figure 8.7 shows the scanning electron microscopy (SEM) image of the $AlSi_{10}Mg$ alloy powder particle distribution, and it can be observed that powder particles are not spherical [1]. The powder particle morphology has a very uneven shape, with many little irregular satellite particles connected to the large powder particles, and other places have seen these irregular structures with little satellite particles [2]. All samples are produced in a melt pool with good metallurgical bonds. We used the high energy (E) input and then created large pores with a low scan speed. The overlapping between molten pools is expected to affect defect formation. Irregularly shaped pores were apparent with more thermal distortion by the high laser of 300 Watts and lower scan of 400 mm/s [5]. The majority of the burring layer was current in the cross-section at a high laser with a 600 mm/s scan speed. It is clear from this that the burring phenomenon is primarily caused by metallurgical porosity defects on the specimen when using a high-powered laser with a fast scan rate [7, 9]. The SLM samples manufactured at T2 show a better microstructure with no visible pores. The laser power energy density is defined as the optimal parameter for obtaining minimum defects, porosity, pores, and cracks for a range of 250 Watts, 500 mm/s, 80 μm, and 30 μm of layer thickness for SLM of $AlSi_{10}Mg$ alloy to achieve a defect-free component.

As built conditions were tested on all samples, the microhardness was higher at T2 at 108.3 HV and lower at T6 Watts at 82 HV due to the high laser with overlap samples. Finally, the hardness and density values are as shown in Figure 8.8, with the highest hardness and density being 108 HV, or 98.6%.

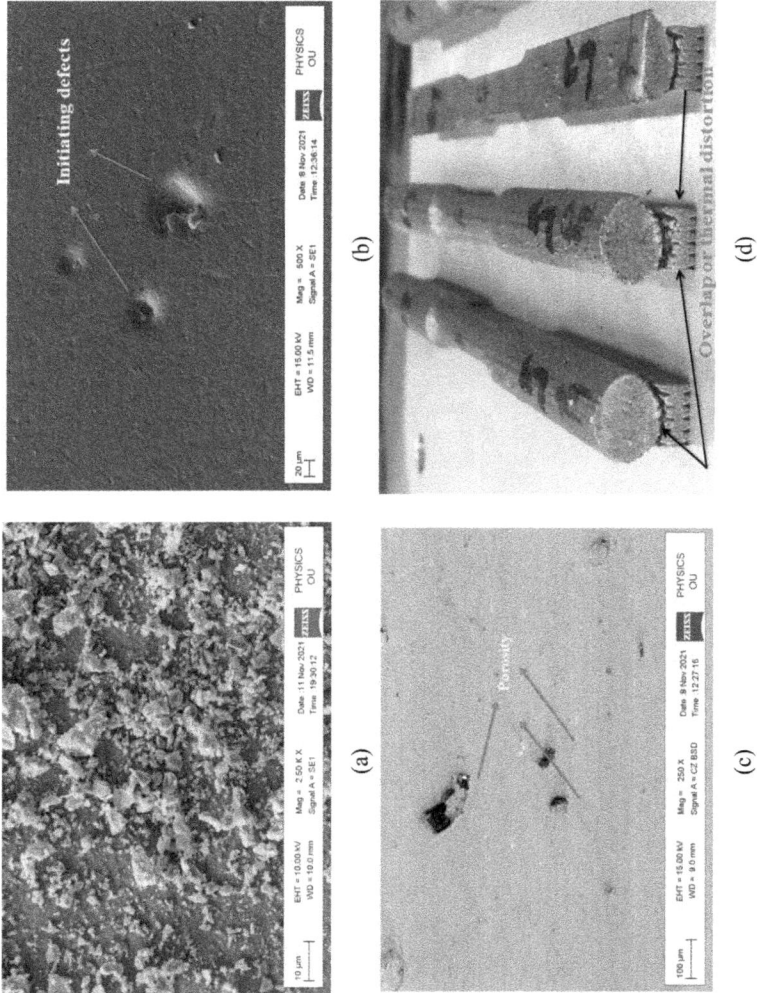

Figure 8.7 The microstructure profile: a) overlap with high 300 Watts laser power b) initiating defect at 300 Watts with 600 mm/s c) porosity at 250 Watts with 400 mm/s and d) fine structure at 250 Watts with 500 mm/s.

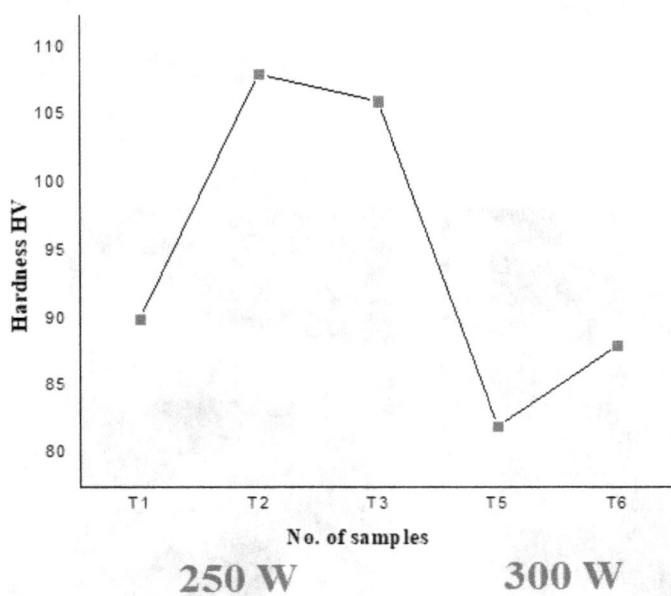

(a)

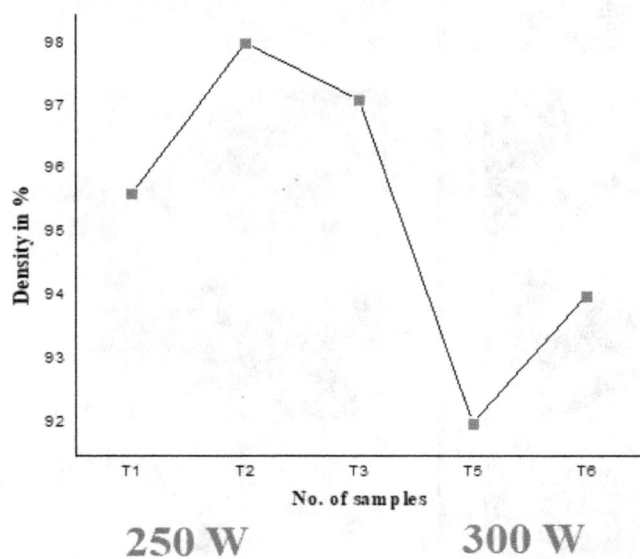

(b)

Figure 8.8 (a) Hardness and (b) Density values.

8.4 CONCLUSIONS

The metallurgical pores, fractures, and isolated areas with overlap of thermal gradient created at high laser power and low scan speed, i.e., keyhole, are presented in this work, and their effects on fatigue strength, density, and hardness of $AlSi_{10}Mg$ alloy are discussed.

- The pores in the overlap boundary are expected to cause early fracture in the samples during fatigue testing.
- The pore overlap and defects were achieved by increasing the hatch distance from 80 to 100 μm (i.e., heat buildup with cool slowly).
- The overlapped and isolated samples have a similar microstructure with little difference in grain sizes. The energy density (E) has an important influence factor on mechanical properties, and defects are formed when the laser energy density is too high or too low (such as defects, porosity, and microcracks). When balling occurs, scan speed is increased, which promotes the capture of powder that has not entirely melted by the laser beam scanning the next layer, resulting in the formation of keyhole pores.
- To avoid defects (pores and cracks) and achieve a fully dense part, low laser power and scan speed should be selected. For that, the best process parameters were obtained when scan speed was 500 mm/s, hatch spacing is 80 μm, and lower power was 250 W with a layer thickness of 30 μm at a fatigue strength of 1.46×10^3, hardness was 108.3 HV, and density is 98% with a laser energy density of 208.3 J/mm^3.

Future work should focus principally on how to eliminate pores and cracks with optimization of process parameters and further improve the fatigue strength, especially in full dense parts.

ACKNOWLEDGEMENT

The author would like to thank the PhD research facilities centre funded by Rashtriya Uchchatar Shiksha Abhiyan (RUSA 2.0) of the government of India at the University College of Engineering (A), Osmania University, Hyderabad, India.

CONTRIBUTIONS OF AUTHORS

Conceptualization: Mudda Nirish , R. Rajendra; Methodology, Mudda Nirish, R. Rajendra, Software, Mudda Nirish; Formal analysis, Mudda Nirish; Investigation, Mudda Nirish, R. Rajendra,; Data curation, Mudda Nirish,

R. Rajendra; Writing—original draft preparation, Mudda Nirish; Writing—review and editing, Mudda Nirish, R. Rajendra; Visualization, Mudda Nirish; Supervision, R. Rajendra.; Funding acquisition, Mudda Nirish. All authors have read and agreed to the published version of the manuscript.

FUNDING

This research received no external funding.

CONFLICT OF INTERESTS

This research received no Conflict of interests.

REFERENCES

[1] Lam LP, Zhang DQ, Liu ZH, Chua CK. Phase analysis and microstructure characterisation of $AlSi_{10}Mg$ parts produced by Selective Laser Melting, *Virtual and Physical Prototyping*, 2015 Oct 2;10(4):207–15.

[2] Kempen K, Thijs L, Yasa E, Badrossamay M, Verheecke W, Kruth JP. Process optimization and microstructural analysis for selective laser melting of $AlSi_{10}Mg$. In *2011 International Solid Freeform Fabrication Symposium 2011 Aug 17*. University of Texas at Austin.

[3] Nirish M, Rajendra R. Suitability of metal additive manufacturing processes for part topology optimization–A comparative study. *Materials Today: Proceedings*, 2020 Jan 1;27:1601–7.

[4] Cook PS, Murphy AB. Simulation of melt pool behaviour during additive manufacturing: Underlying physics and progress. *Additive Manufacturing*, 2020 Jan 1;31:100909.

[5] Ngnekou JN, Nadot Y, Henaff G, Nicolai J, Ridosz L. Influence of defect size on the fatigue resistance of $AlSi_{10}Mg$ alloy elaborated by selective laser melting (SLM). *Procedia Structural Integrity*, 2017 Jan 1;7:75–83.

[6] Nirish M, Rajendra R. Effect of heat treatment on wear characterization of $AlSi_{10}Mg$ alloy manufactured by selective laser melting. *Balance*, 2022;9:11.

[7] Han Q, Setchi R, Lacan F, Gu D, Evans SL. Selective laser melting of advanced $Al\text{-}Al_2O_3$ nanocomposites: simulation, microstructure and mechanical properties. *Materials Science and Engineering: A*, 2017 Jun 20;698:162–73.

[8] Mudda N., Rajendra R. Additive simulation and process parameter optimization for wear characterization development by selective laser melting of $AlSi_{10}Mg$ alloy. *Journal of Characterization*, 2(2):103–116, ISSN: 2757-9166.

[9] Sutton AT, Kriewall CS, Leu MC, Newkirk JW. Powder characterisation techniques and effects of powder characteristics on part properties in powder-bed fusion processes. *Virtual and Physical Prototyping*, 2017 Jan 2;12(1):3–29.

[10] Rometsch P, Jia Q, Yang KV, Wu X. Aluminum alloys for selective laser melting–towards improved performance. In *Additive Manufacturing for the Aerospace Industry*, 2019 Jan 1 (pp. 301–325). Elsevier.

[11] Wang P, Lei H, Zhu X, Chen H, Fang D. Influence of manufacturing geometric defects on the mechanical properties of AlSi$_{10}$Mg alloy fabricated by selective laser melting. *Journal of Alloys and Compounds*, 2019 Jun 15;789: 852–9.

[12] Nirish M, Rajendra R. Heat treatment effect on the mechanical properties of AlSi$_{10}$Mg produced by selective laser melting. *Journal of Mechanical Engineering Research and Development*, 2022, 45:19–28.

[13] Beretta S, Gargourimotlagh M, Foletti S, Du Plessis A, Riccio M. Fatigue strength assessment of "as built" AlSi$_{10}$Mg manufactured by SLM with different build orientations. *International Journal of Fatigue*, 2020 Oct 1;139: 105737.

[14] Ch SR, Raja A, Jayaganthan R, Vasa NJ, Raghunandan M. Study on the fatigue behaviour of selective laser melted AlSi$_{10}$Mg alloy. *Materials Science and Engineering: A*, 2020 Apr 20;781:139180.

[15] Uzan NE, Shneck R, Yeheskel O, Frage N. High-temperature mechanical properties of AlSi$_{10}$Mg specimens fabricated by additive manufacturing using selective laser melting technologies (AM-SLM). *Additive Manufacturing*, 2018;24:257–63.

[16] Nirish M, Rajendra R. Optimization of process parameter and additive simulation for fatigue strength development by selective laser melting of AlSi$_{10}$Mg alloy. *International Journal of Mechanical Engineering*, 2022;7(2):3795–3802. ISSN: 0974-5823.

[17] Xu ZW, Wang Q, Wang XS, Tan CH, Guo MH, Gao PB. High cycle fatigue performance of AlSi$_{10}$Mg alloy produced by selective laser melting. *Mechanics of Materials*, 2020 Sep 1;148:103499.

[18] Bai P, Huo P, Kang T, Zhao Z, Du W, Liang M, Li Y, Liao H, Liu Y. Failure analysis of the tree column structures type AlSi$_{10}$Mg alloy branches manufactured by selective laser melting. *Materials*, 2020 Jan;13(18):3969.

[19] Qian G, Jian Z, Qian Y, Pan X, Ma X, Hong Y. Very-high-cycle fatigue behavior of AlSi$_{10}$Mg manufactured by selective laser melting: Effect of build orientation and mean stress. *International Journal of Fatigue*, 2020 Sep 1;138:105696.

[20] Zhang W, Hu Y, Ma X, Qian G, Zhang J, Yang Z, Berto F. Very-high-cycle fatigue behavior of AlSi$_{10}$Mg manufactured by selected laser melting: Crystal plasticity modeling. *International Journal of Fatigue*, 2021 Apr 1;145:106109.

[21] Nirish M, Rajendra R. Optimization process parameter on wear characterization of Al6061 and AlSi$_{10}$Mg alloy manufactured by selective laser melting. *International Journal of 3D Printing and Additive Manufacturing*, 2022; 1(1):1–10.

[22] Santos LM, Ferreira JA, Jesus JS, Costa JM, Capela C. Fatigue behaviour of selective laser melting steel components. *Theoretical and Applied Fracture Mechanics*, 2016 Oct 1;85:9–15.

[23] Uzan NE, Ramati S, Shneck R, Frage N, Yeheskel O. On the effect of shot-peening on fatigue resistance of AlSi$_{10}$Mg specimens fabricated by additive manufacturing using selective laser melting (AM-SLM). *Additive Manufacturing*, 2018 May 1;21:458–64.

[24] Jian ZM, Qian GA, Paolino DS, Tridello A, Berto F, Hong YS. Crack initiation behavior and fatigue performance up to very-high-cycle regime of AlSi$_{10}$Mg fabricated by selective laser melting with two powder sizes. *International Journal of Fatigue*, 2021 Feb 1;143:106013.

[25] Brandl E, Heckenberger U, Holzinger V, Buchbinder D. Additive manufactured AlSi$_{10}$Mg samples using selective laser Melting (SLM): Microstructure, high cycle fatigue, and fracture behavior. *Materials & Design*, 2012 Feb 1;34:159–69.

[26] Hamidi Nasab M, Giussani A, Gastaldi D, Tirelli V, Vedani M. Effect of surface and subsurface defects on fatigue behavior of AlSi$_{10}$Mg alloy processed by laser powder bed fusion (L-PBF). *Metals*, 2019 Oct; 9(10):1063.

Chapter 9

Role of artificial intelligence in the apparel industry in the context of Industry 4.0 and Industry 5.0

Tulasi B.
CHRIST (Deemed to be University), Bangalore, Karnataka, India

Sudhir Karanam
Reliance Retail, Bangalore, Karnataka, India

CONTENTS

9.1 INTRODUCTION

Transformations led by industrial revolutions have impacted many societal aspects such as economy, transportation, and healthcare. They have led to disruptive innovations and have had overwhelming impact on the daily life of people. From the 18th century onward, industrial revolutions have always led to the merger of technology into various industries. The first three of these revolutions were to be classified as mechanical, electrical and automotive. The Fourth Industry Revolution (or Industry 4.0) originated in the year 2011 from a German project and one of its prime initiative is the so-called smart factory [1]. The smart factory is equipped with intelligent sensors and embedded software which communicate to provide new ways of production, better decision-making and optimizations. This is also labeled a "Cyber-Physical System" [2]. Artificial intelligence (AI) is one of the main elements in the technology basket associated with Industry 4.0,

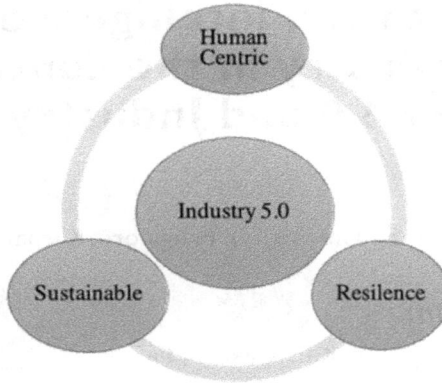

Figure 9.1 Focal points of Industry 5.0.

whose twin main goals are optimum performance and increased efficiency. While value chain efficiency has been the primary focus, creativity is an aspect which is now being looked upon.

The next development in industrialization is now considered to be mass customization and personalization to the maximum level. This is the focus of Industry 5.0. In 2021, the European Commission formally introduced "Industry 5.0" after detailed discussions and deliberations. Its key aspects are "Sustainability"," Resilience" and "Human-Centric," as depicted in Figure 9.1. The proposed Industry 5.0 emphasises the collaboration of precise automation with human creativity and critical thinking skills. Its intent is the creation of high-value jobs rather than the replacement of monotonous and repetitive tasks by automation in Industry 4.0. Intelligent healthcare, supply chain, manufacturing process, 6G and beyond are just a few of the potential applications of Industry 5.0. The requirement of intellectual professionals to work in collaboration with automated technology is also seen as one of the driving forces of this new revolution. The enabling technologies of Industry 5.0 have been identified as follows: (1) Bio-inspired technologies, (2) Energy efficiency and autonomy, (3) Human–machine interaction, (4) Digital twins and simulation, (5) Data transmission storage and analysis, and (6) Artificial intelligence (AI). The blend of automatization and cognitive intelligence can be obtained through AI; "cobots" (or collaborative robots) is a term coined in Industry 5.0. These collaborative robots are designed to have simple and intuitive interactions with humans to ensure the right blend of accuracy and creativity in all manufacturing processes.

9.1.1 Industry 4.0, Industry 5.0 and the apparel industry

Adopting the automation and data trends of Industry 4.0, the apparel industry has developed the *"Smart Apparel Factory"* [3, 20]. Various digital technologies, such as cloud computing and blockchain, have had tremendous impacts on the production life cycle of the apparel industry. The factory has moved towards Cyber-Physical systems for more efficient and effective approach for production. AI techniques have been used to handle various decision-making processes in the apparel and fashion industries such as production scheduling, sales forecasting, and fashion recommendations. AI techniques are broadly classified into two categories: "Symbolic AI" and "Computational Intelligence". These could equally be classified as "Knowledge-based systems" and "Evolutionary computation techniques", respectively. These techniques have been used to obtain improved productivity through the interconnectedness of machines. The application of AI to the apparel industry has led to the development of enhanced business models [3]. Areas of designing, merchandising, personalization to customer requirements, and trend forecasting have also been tremendously impacted through the application of AI. Generative adversarial networks (GANs), augmented reality (AR), virtual reality (VR) and Neural Machine Translation (NMT) algorithms have all been adopted in the above-mentioned areas of the apparel industry.

Product Life Cycle Management (PLM), a software application has also been adopted as part of the apparel industry in line with Industry 4.0. PLM covers the process from the conception of the design to the final stage of sales in the retail store. It is a tool which provides the apparel industry with an easy way to handle all of the operations involved. Two of the major benefits of PLM are: (i) the optimization of time involved in various processes and (ii) cost effectiveness. Studies indicate that the time required for the production life cycle has been halved and that revenues have improved by 13.4% through the use of the software application. It offers businesses a simple and clean way to enhance their production. Although PLM has been effective, its amalgamation with 3D technology would have a profound impact on the apparel industry. It has been noted that the use of 3D technology has helped designers to create a smooth and efficient digital workflow [8]. The use of technology in the fashion and apparel industry has paved the way for digital fashion and a move away from mass production processes. 3D technology is widely seen as the much-required platform to accelerate the penetration of AI into the apparel industry. The use of 3D modelling helps to reduce production costs.

The digital fashion, which is the label given to the entire process for garment creation to become digital, ensures the correction of minor mistakes in the design to be handled in an economical manner. 3D technology then helps by avoiding the use of non-compactible materials, thereby thus leading to a sustainable process, which is the stated goal of Industry 5.0. It also leads to an environment-conscious approach of production which is a vital need considering that the apparel industry is a major stakeholder in the waste generated globally. Data-backed, the "Digital Opinion" given by an AI application to a designer would not only be neutral and unbiased, but would also be personalized as per the client's needs [6]. Human creativity coupled with machine learning (ML), a division of AI, can design for present and future, thereby providing a "dream garment". This is a major technological leap in the apparel industry, which moves towards the "Human-Centric" goal of Industry 5.0.

9.1.2 Digital Intelligence – the first step towards artificial intelligence in the apparel industry

The Integration of artificial intelligence (AI) into the apparel industry would require rapid changes at multiple levels—in terms of policy and individuals. Industry 4.0 and 5.0 require an interdisciplinary work environment [5]. The workforce should possess digital skills and competencies to interact with the digital environment and media [9]. This indicates that the workforce should be digitally intelligent, an ability to think and adapt in the ever-changing digital environment. "DQ Institute", a global think-tank and social enterprise, describes Digital Intelligence (DI) as an important skill to thrive in this digital era [12]. The competencies proposed by the institute emphasises the eight major elements: Digital Identity, Digital Rights, Digital Literacy, Digital Security, Digital Safety, Digital Use, Digital Emotional Intelligence, and Digital Communication. Figure 9.2 depicts the global standards of competency IEEE 3527.1.

Each of the eight elements is further subdivided into three sections. The most important of these in the context of apparel industry is Digital Literacy, which subdivides into Data and AI Literacy; Content Creation and Computational Literacy;, and Media and Information Literacy. The workforce of the industry should be competent to process, analyse the

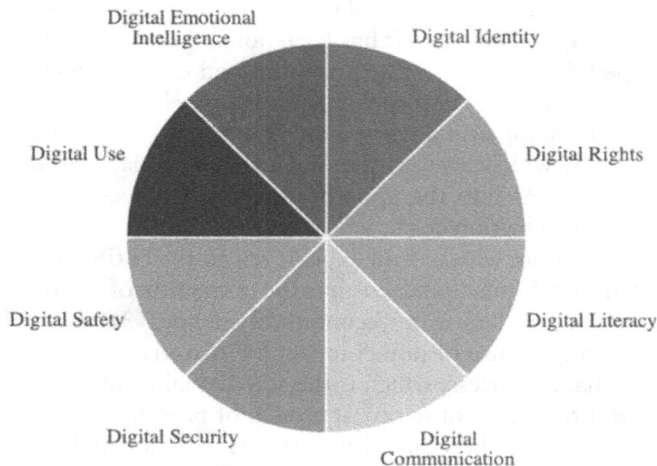

Figure 9.2 Global standard IEEE 3527.1 – DQ institute.

mathematical models and understand computational logic, thereby enabling them to take data-driven relevant decisions [14]. Understanding the creation and statistical analysis of data, the use of AI algorithms such as machine learning and eep learning helps in understanding the patterns and facilitating decision-making. This emphasises that digital competency is an essential skill in Industry 4.0 and 5.0.

9.2 AI IN PRODUCTION PROCESS OF APPAREL INDUSTRY

Two areas of concerns in the production processes of apparel industry are the enhancement of the efficiency of production process and also cost-effectiveness [11]. Both of these can be addressed by blending AI into the production process. AI can provide enhanced solutions through its heuristic algorithms. A typical production in apparel industry would have: (1) Ideation (2) Design Conceptualization (3) Apparel manufacturing (4) Retailing. Figure 9.3 depicts the major stages of production. AI can play a vital role in all stages of production and it also helps in providing information flow from retailing to ideation for trend forecasting. In the ideation stage, the designer would conceptualize the theme depending on the forecasting of trends in colours and fabrics, followed by the designing of the apparel. Apparel manufacturing includes fabric spreading, cutting, bundling, sewing, inspection and packaging.

The final stage is retailing. At this stage the product is sold to the customers, who, in turn, can provide data to stimulate the ideation process. The initial stage of ideation is one of the important phases of the production process and the application of AI techniques at this point can be crucial.

9.2.1 AI as a part of design

Designers outline themes which would be futuristic trends of the industry. Digitisation has helped all sectors of the market in obtaining much-needed data for forecasting. The apparel and fashion industry also benefits strongly

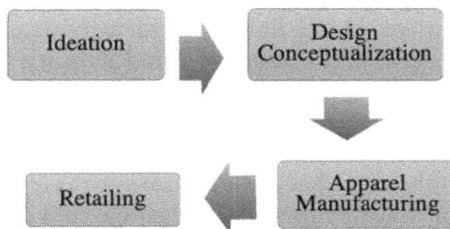

Figure 9.3 Major stages of production in the apparel industry.

from the digital trails of data available. This data, when analysed, leads to knowledge which impacts the decision-making of designers [4]. AI is the new ally for the designers and can also be considered as the new advisor. The ever-changing trends and causes for the change can be easily tracked and reasoned with the help of AI algorithms. AI-driven applications such as "ImaGenie" from Stylumia Labs analyse large datasets of images and the textual attributes of fashion styles across the digital platforms; it also predicts the pioneer design and trends [13]. Similarly, an AI tool developed by a group of professionals in Amazon analyses images in order to generate a new design by itself. Supplementing this is another AI application which analyses an image to predict whether it is a trendy design. H&M, the Swedish fashion brand, has proactively employed AI to predict fashion trends and customer behaviours, leading to sustainable decisions reducing the wastage usually incurred. The digital opinion about the designs is another very important utility of the AI application in the design process.

Product imaginary is an essential requirement in the industry and has always impacted the consumer market. It provides the final look which is essential for a customer to relate to the product, this leads to increased purchase. AI-generated 3D product imagery has facilitated the provision of virtual product imagery with just a "sketch" of a garment. The AI-generated clothing on AI-generated models reduces the budget and the time consumed when compared with the physical product imaginary [10]. This enhances the timeline to deliver the products and is also a sustainable process.

9.2.2 AI as part of apparel manufacturing

Fabric is the primary input to the apparel industry. The quality of fabric is checked once it is received from the supply chain system. The fabric is then spread for cutting as per the design requirements. Several pieces of equipment are utilised to cut the fabric which is then sent to the sewing floor. Skilled operators stich the fabric and then the finished apparel undergoes a quality check before packaging. Due to the ever-changing demands of the customer and to adhere to the personalization of the product, the manual process of manufacturing tends to be both error-prone and tedious. Current applications of AI are limited to identifying and grading fabric based on colour, length, uniformity ratio and tenacity. But areas like virtual modelling of yarn, prediction of yarn unevenness, and the detection of wrinkles on the fabric is where AI would connect and provide the require solution [16]. The cutting of the fabric also generates overheads in terms of the wastage of fabric; AI-backed applications have proved effective in this respect and one study suggests that wastage could be reduced to only 1.7% [21]. Applications provide real-time visibility of the cutting process, thereby enabling the correct estimate of the fabric. This leads to enhanced productivity and cost-effectiveness. Similarly, automated "Sewbots" use machine vision to analyse the fabric and adjust as per the design and the completion of the sewing of the apparel [15, 17, 23]. These applications indicate the presence of AI across the entire life cycle of apparel manufacturing.

9.3 AI AND CUSTOMER EXPERIENCE

Retail in the fashion and apparel industry has overwhelmingly moved into the digital environment. AI has impacted the e-commerce sector through personalization and effective customer interaction. The overall customer experience can also be enhanced by the use of AI tools. With the integration of AI, augmented reality and virtual avatars, the digital shopping experience has been improved. Digital Customer Satisfaction (DCX) is term given for the sum total experience a customer has with a brand or website [18]. Digital customer satisfaction can be considered to be a subset of Customer Experience (CX), which encompasses in-store interactions and other physical world interactions. CX focuses on building trust, human connection and empathy and DCX does the same in the digital world. It is important that customers have seamless experience in both the physical and the digital environment. Customer data is important to personalize the digital experience and to ensure relevant content and experiences are provided to a customer. It is essential to ensure that all customers have outstanding DCX as this leads to higher customer retention, reduced customer churns and greater brand equity. Machine learning and other AI techniques provide Digital Experience Platforms (DXP) that collect, analyse and interpret data to enhance DCX.

9.3.1 Hyper-personalization and AI

Dynamically evolving customer expectations require Hyper-Personalization to create meaningful connections with customers using DXP. Hyper-Personalization is considered as the most relevant means of communication for a customer. It ensures that all the requirements of a customer are considered. According to Salesforce's 2020 report, more than 70% of customers want brands to have a better understanding of their needs [19]. Hence, the requirement for Hyper-Personalization—beyond personalization. Hyper-Personalization uses real-time data, AI, ML, and Predictive Analytics to understand individual customers' wants and needs. It provides specific actions and approaches in connecting to each customer in a more relevant manner. The experiences are tailor-made to individual needs and thus enhance the personalization. Hyper-Personalization requires the modeling of different, varied behaviours and apply algorithms on real-time data to decide on the optimal timeline in order to communicate to the customer.

It anticipates the needs of the customers accurately by constant engagement. The process of Hyper-Personalization consists of four major elements:

- Data: Optimizing the customer data to create personalized strategies for enhancing the Hyper-Personalized customer journey.
- Content: The creation of specific content for each individual and identifying the method of delivery so as to convert a visitor as a customer.
- Deliver: Ensure that consistent individualised experience is available at the right time as per the customer's personal preference.
- Unify: Combine all the above elements using DXP to automate and apply required algorithms for optimizing the experience.

Amazon, Airbnb and Reebok are just some of the pioneers in adopting Hyper-Personalization [7]. Though customers enjoy personalization, privacy is a differentiator. It is essential that the customer is comfortable with the leverage of data and that transparency is maintained in all communications.

9.4 CONCLUSION

Industrial revolutions have always significantly impacted on every aspect of everyday life. At present, Industry 4.0 is revolutionizing the way industries manufacture and distribute their products. The integration of technologies such as artificial intelligence (AI), the Internet of Things (IoT), cloud computing, and analytics have laid foundation to the "Smart Factory". These technologies, and specifically AI, have accelerated automation and optimised every process across individual industries. Industry 5.0 is moving beyond efficiency and production goals, and is focused on making the process more human- or society-centric [22, 24]. Accordingly, sustainability is one of its main focal points. The apparel industry is a labour-intensive industry and also plays a significant role in the global economic landscape. In adopting the goals of Industry 4.0 and 5.0 into the apparel industry, AI would play an important role. This chapter has tried to highlight the multifunctional role AI plays in the apparel industry. The various technologies mentioned are sources of value creation and are sustainable in the long run. Future research should definitely lead to a deeper and balanced implementation of the advanced AI technologies.

REFERENCES

[1] Vogel-Heuser, B., & Hess, D. (2016). Guest editorial Industry 4.0–prerequisites and visions. *IEEE Transactions on Automation Science and Engineering*, 13(2), 411–413.
[2] Bauernhans, T., Vogel-Heuser, B., & ten Hompel, M. In: Allgemeine Grundlagen, editor, *Handbuch Industrie 4.0 Bd.4*. Springer; 2017. ISBN: 978-3-662-53254-53256.

[3] Nayak, R., & Padhye, R. (2018). Artificial intelligence and its application in the apparel industry. In *Automation in Garment Manufacturing*. Amsterdam, The Netherlands: Elsevier, pp. 109–138.

[4] Acharya, A., Singh, S. K., Pereira, V., & Singh, P. (2018). Big data, knowledge co-creation and decision making in fashion industry. *International Journal of Information Management*, 42, 90–101.

[5] Moore, M. What is Industry 4.0? Everything you need to know. *TechRadar*. 27 May 2020.

[6] Müller, J. Enabling Technologies for Industry 5.0, Results of a workshop with Europe's technology leaders, Directorate-General for Research and Innovation (European Commission), Published: 2020-11-26.

[7] Adejeest, D.-A. Amazon's U.S. Market Share of Clothing Soars to 14.6 Percent. 2022. Available online: https://fashionunited.com/

[8] Backs, S., Jahnke, H., Lüpke, L., Stücken, M., & Stummer, C. (2021). Traditional versus fast fashion supply chains in the apparel industry: An agent-based simulation approach. *Annals of Operations Research*, 305, 487–512.

[9] Kampakaki, E., & Papahristou, E. (n.d.). Digital Intelligence as Prerequisite of Artificial Intelligence's Integration in the Clothing Industry 4.0 SETN2020 Workshops.

[10] Giri, C., Jain, S., Zeng, X., & Bruniaux, P. (2019). A detailed review of artificial intelligence applied in the fashion and apparel industry. *IEEE Access* 7, 95376–95396. https://doi.org/10.1109/ACCESS.2019.2928979

[11] Li, Y., Dai, J., & Cui, L. (2020). The impact of digital technologies on economic and environmental performance in the context of industry 4.0: A moderated mediation model. *International Journal of Production Economics*, 229, 107777.

[12] DQ Framework|DQ Institute. (2020). https://www.dqinstitute.org/dq-framework/#digital_intelligence

[13] Gao, Y., Li, X., Wang, X.V., Wang, L., & Gao, L. (2021). A review on recent advances in vision-based defect recognition towards industrial intelligence. *International Journal of Industrial and Manufacturing Systems Engineering*, 62, 753–766.

[14] Vista, A. (2020). Data-driven identification of skills for the future: 21st-century skills for the 21st-century workforce. *SAGE Open*, 10, 2158244020915904.

[15] Lindström, J., Kyösti, P., Birk, W., & Lejon, E. (2020). An initial model for zero defect manufacturing. *Apply Science (Switzerland)*, 10, 4570.

[16] Alemão, D., Dionisio Rocha, A., Barata, J., Alemão, D., & Rocha, A.D. (2021). Smart manufacturing scheduling approaches-systematic review and future directions. *Applied Sciences*, 11(5), 2186.

[17] Zhou, L., Zhang, L., & Horn, B.K.P. (2020). Deep reinforcement learning-based dynamic scheduling in smart manufacturing. *Procedia Cirp*, 93(January), 383–388.

[18] Barni, A., Pietraroia, D., Zust, S., West, S., & Stoll, O. (2020). Digital twin based optimization of a manufacturing execution system to handle high degrees of customer specifications. *Journal of Manufacturing and Materials Processing*, 4(December 4), 109.

[19] Giri, C., Thomassey, S., & Zeng, X. (2019). Customer Analytics in Fashion Retail Industry. In *Functional Textiles and Clothing*; Singapore: Springer, pp. 349–361.

[20] Udayangani, J., Karunanayaka, I., & Abeysooriya, R. (2019). Industry 4.0 Elements and Analytics for Garment Assembly Production Lines. *2019 Moratuwa Engineering Research Conference (MERCon)*, 745–750, doi: 10.1109/MERCon.2019.8818882.

[21] Jana, P. (2018). Automation in Sewing Technology. *Automation in Garment Manufacturing*, pp. 199–236. Elsevier.

[22] Johri, P., Singh, J. N., Sharma, A., & Rastogi, D. (2021). Sustainability of Coexistence of Humans and Machines: An Evolution of Industry 5.0 from Industry 4.0. *2021 10th International Conference on System Modeling & Advancement in Research Trends (SMART)*, pp. 410–414, doi: 10.1109/SMART52563.2021.9676275.

[23] Nahavandi, S. (2019). Industry 5.0—A human-centric solution. *Sustainability*, 11(16), 4371.

[24] Iqbal, M., Lee, C. K. M., & Ren, J. Z. (2022). Industry 5.0: From Manufacturing Industry to Sustainable Society. *2022 IEEE International Conference on Industrial Engineering and Engineering Management (IEEM)*, pp. 1416–1421, doi: 10.1109/IEEM55944.2022.9989705.

Chapter 10

Swarm intelligence-based automotive manufacturing

Hiranmoy Samanta and Kamal Golui

Gargi Memorial Institute of Technology, Baruipur, Kolkata, India

CONTENTS

10.1 INTRODUCTION

Aggravations in assembling frameworks are spontaneous changes, which are arranged by recuperating time, including brief time frame, long time and failure to recuperate; or characterized by upset levels, including inward and outside unsettling influences. Interior unsettling influences occur at the control frameworks, creation types of gear, material dealing with and work. Outside unsettling influences occur at the cycle steps connected with the orders, stocks and providers. At present, the traditional assembling

DOI: 10.1201/9781003257714-10

frameworks ought to be halted to fix the damaged machine, which reduces their usefulness and builds the vacation of the assembling frameworks. The variation of an assembling framework to aggravations is the capacities to answer quickly and to recuperate independently, which keeps the assembling framework running and avoids it being shut down completely. Reconfiguration and independence are measures to adjust to the unsettling influences in an efficient manner. Reconfiguration revises and rebuilds the assembling assets that require a rescheduling and reconfigurable capacity of assembling frameworks. The customary assembling frameworks can't look with the unsettling influences by their unbending construction. These frameworks should be halted when the aggravations take place. The chapter presents an Autonomous Manufacturing System in light of Swarm of Cognitive Agents (AMS-SCA) [1] to adjust to the difficulties. In this structure, the assembling framework is considered as a multitude of mental specialists where tools, machines, robots, and carriers are constrained by the relating mental specialists. In short, the objectives of this work were:

- To formulate a model using IoT, RFID, AMS-SCA, IMS that would permit the control framework to make a move when the aggravation occurs and to continue to work rather than halting the assembling framework totally.
- To prepare elements in the assembling framework with the direction and self-controlling capacities.

10.2 LITERATURE REVIEW

A wide range of studies have been conducted into swarm-based manufacturing. Cheraghi et al. [2] presented a study on the emerging field of swarm robotics [2–5]. They describe the technology to be a mimic of the natural swarms in order to increase the effectiveness. Important features and applications of this approach have also been discussed. Ismail and Hamami [3] worked on a target-searching problem using Swarm Robotic Strategies. They have carried out a systematic review of different techniques used to solve this problem and then filtered the extracted data with two databases, namely Scopus and Web of Science. Cai and Sharma [6] have built an algorithm model using V3CFOA, evaluated the collected data set and then compared it with the details obtained using other algorithms. It has been found that this model and algorithm helped to predicting the prevalence of a crop pest with higher accuracy, which results in a more stable output of crops. Guo and Garcia [7] have tended to enter innovation issues in the space of information procurement and pre-processing, digital actual combination, information extraction and sharing, and gadget execution self-streamlining. The proposed approach is also exceptionally useful in further developing the handling productivity of the reconfigured creation line. Monga et al. [8]

have worked on different intelligence techniques based on swarm technology. This analysis comprises four stages: planning, shortlisting, extraction and execution. It is found that all these methods can be used to solve different problems. A proper balancing of this high-level problem-independent algorithmic framework is required to obtain proper performance. Schranz et al. [9] have dealt with classifying a multitude of different ways of behaving into the spatial association, route, navigation, and various. Then this grouping is applied to existing multitude applications. An overview is carried out based on this order. Found swarm applications are rarely used. The reasons have been made sense of and the issues have been recognized. Kaur and Kumar [10] reviewed the application of swarm intelligence in different computing tasks. They concentrated on Bat calculation, Firefly calculation, Lion enhancement calculation, Chicken multitude improvement calculation, Social Spider Algorithm and Spider Monkey advancement calculation. In addition, they have likewise done a complete report on cloud figuring, fog registering and edge processing utilizing swarm knowledge. They showed that in each case there have been improvements after the incorporation of swarm intelligence. Ramanan et al. [11] worked on improving the finished product using intelligent particle swarm optimization methods. They varied different parameters as discharge current, servo speed rate, etc. to check its effect on the output processes. A numerical model has been laid out to do the examination of difference and finish up the ideal machining conditions utilizing the molecule swarm enhancement method. Sibalija [12] have reviewed different research works on the optimization of complex manufacturing processes using particle swarm technology. Other optimising algorithms have been compared with the results of particle swarm optimization [6, 11–17] in different manufacturing processes considering different parameters. The review analysis of optimising the manufacturing process of the past decade has been presented in this paper. Qin et al. [13] have examined design-relevant features with respect to energy modelling. A deep learning-driven particle swarm optimization has been proposed for the optimization of energy utility. The research work has been validated with data collected from real-world additive manufacturing industries. Guerreiro et al. [18] have discussed about the problems faced by the industries in shifting paradigms and the role of Big Data Analytics in it. They have discussed an instance of unloading car batteries and their fitting where swarm architecture has been used in the processing of data in the data-driven architecture. Wang and Feng [19] have discussed how the particle swarm optimization method could be used to measure the green degree (the degree of safeness) of equipment manufacturing industry. They found that swarm optimization is much handier than the previously used methods. Chamanbaz et al. [20] have worked on enabling swarm technology in multi-robotic systems. They designed by integrating the software and hardware in such a way as it could be easily ported across various robotic platforms. They implemented in differently distributed multi-robotic systems. They found that this design facilitates different new swarm behaviors. Navarro and Matía [4] have given an outline of multitude mechanical technology in this paper. The properties and its qualities have been done alongside an examination of general multi-robot framework. The future promising applications, along with the issues to defeat to contact them have been made sense of and investigated. Park and Tran [1] have proposed an autonomous manufacturing system using swarm cognitive agents. In this model each and every component is controlled by cognitive agents and are capable of responding to the problems arising in the process of manufacturing. Gandolfo Dominic [21] have represented a theoretical framework of the Holonic Production System and the ways to overcome problems arising in conventional methods of production. It has been said in the paper that the capability of the Holonic Production System to adjust and respond to changes in the business climate while having the option to keep up with fundamental collaborations and coordination through the holonic structure where utilitarian creation units are at the same time independent and agreeable. Navalertporn and Afzulpurkar [14] have proposed a coordinated enhancement strategy using counterfeit brain organizations and bidirectional multitude improvement. This strategy has been applied to tile assembling and observed that it is equipped for taking care of intricate interaction boundary plan issue. Bannat et al. [22] presented the overall standards of independence and the proposed ideas, strategies and advancements to acknowledge the mental preparation, mental control and mental activity of creation frameworks. Labour direction frameworks for manual gathering with naturally and situationally subordinate set off ways on state-based diagrams are depicted in this paper. Wu [15] have proposed a model of interest determining joining the part support vector machine and multitude enhancement. They additionally contrasted their outcomes and the determining after effects of cross-breed PSOv-SVM and different strategies and observed that their proposed strategy is superior to the others. Durán et al. [16] have addressed the problem of the construction of a manufacturing cell using particle swarm optimization. The algorithm proposed here is normally used in data-mining processes. It is found that optimal results have been obtained in all cases using this process. Shea et al. [23] have presented another methodology and structure for an independent plan-to-creation framework that coordinates mental abilities, for example, thinking from information models and independent preparation, and implants these in the actual machines to manufacture altered parts consequently. The system coordinates into a typical interaction programmed workpiece choice utilizing a cosmology, generative CNC machining arranging utilizing shape syntaxes and mechanized installation configuration, in view of a clever adaptable apparatus gadget equipment. Leitao [24] have overviewed the writing in assembling control frameworks utilizing dispersed man-made brainpower strategies, in particular multi-specialist frameworks and holonic fabricating frameworks standards. The paper have additionally examined the explanations behind the frail reception of these methodologies by industry and brings up the difficulties and exploration open doors for what's in store. Wang et al. [25] have introduced an incorporated way to deal with handle the interruption of machine breakdown for the unique responsive booking issue. The proposed rescheduling approach is accomplished in an iterative way by investigating the cycle plan arrangement space of the positions in question. An integrator module is utilized to connect the interaction arranging module and planning module. A heuristic rule is proposed to change arrangement space as far as limitations, which are taken care of back to the cycle arranging module. Anghinolfi et al. [26] have introduced a search cycle to confront the NP-hard single machine all out weighted lateness planning issue in presence of grouping subordinate arrangement times. This cycle is called Discrete Molecule Multitude Enhancement. They compared their results with the other's research works and found that the effectivity of their DPSO swarm intelligence was the best. Christo and Cardeira [27] have categorised Intelligent Manufacturing System as Fractal, Bionic and Holonic. They are said to be the future of manufacturing industry. The principles for constructing an intelligent manufacturing system has been discussed in this paper. Zhou et al. [28] have introduced a plan technique for independent and helpful FMS control frameworks in light of the most recent specialist guidelines of the establishment clever actual specialist to make control elements of adaptable assembling frameworks having qualities of independence, participation, vigour, modularization, reconfiguration and practicality. Monostori et al. [29] have presented programming specialists and multi-specialist frameworks and, through an exhaustive study, their potential assembling applications are illustrated. Valckenaers et al. [30] have introduced the plan of a holonic fabricating execution framework. The plan is a launch of the Product-Resource-Order-Staff Architecture (PROSA) reference design increased with coordination and control components propelled by regular frameworks. The plan has been applied to a modern experiment and the consequences of this contextual investigation have likewise been talked about in the paper. Gerardo Beni [5] have defined the swarm technology, optimization and swarm robotics and its utilisation in this paper. Vieira et al. [31] have introduced suitable definitions for uses of rescheduling fabricating frameworks and portrayed a system for understanding rescheduling techniques and strategies. The impact of rescheduling on the presentation of assembling framework have likewise been talked about. Leitao and Restivo [32] have introduced a coordinated and versatile assembling control design that tends to the requirement for the quick response to aggravations at the shopfloor level, expanding the readiness and adaptability of the undertaking, when it works in unstable conditions. The proposed design presents a versatile control that adjusts powerfully between a more concentrated structure and a more decentralized one, permitting consolidating the worldwide creation streamlining with deft response to unforeseen unsettling influences. Monostori [33] have examined about how design acknowledgment strategies, master frameworks, counterfeit brain organizations, fluffy frameworks and half-and-half computerized reasoning methods could be applied in assembling frameworks. The main strides of this cycle and a few new outcomes with extraordinary accentuation on half-breed AI and multi-methodology AI approaches have been presented. Van Brussel et al. [30] have worked on Holonic Manufacturing Systems (HMS). The HMS paradigm have been discussed broadly. The properties of a complex adaptive system have also been discussed. Examples have been demonstrated to explain the designing of distributed manufacturing systems and its components.

10.3 RESEARCH METHODOLOGY

This section will outline the research methodology adopted to perform the objectives as mentioned in section 10.1. In short, the research methodology is as follows:

- Various writings and the examination work have been contemplated on Intelligent Manufacturing System.
- The fundamental spotlight on Supply Chain the executives, IMS, IoT, RFID.
- Advancement of the RFID-based framework with the guide of IoT and distributed computing.

10.4 COGNITIVE AGENTS

Cognitive agents or the intellectual specialists are either organic elements including humans and creatures or fake substances like robots and programming specialists. Figure 10.1 shows the cognitive system of manufacturing.

In this case, a cognitive agent is a PC program that uses BDI technology to give professionals false intellectual abilities. Subsequently, the specialist plays out the intellectual exercises which copy the intellectual practices of humans. Intellectual exercises play out a circle of three stages: discernment, thinking and execution. The intellectual specialist has all attributes of the customary specialist, including the independence, social capacity, reactivity, and supportive of animation. The independence is a capacity of the specialist to accomplish its objectives with next to no support from different specialists. On the other hand, the cooperation of experts to achieve the global goals of the framework is the social ability of experts. Reactivity, which relies on the relationship between insight and activity, is the professional's ability to respond to changes in nature. The supportive of animation of specialists is a capacity to communicate the objective coordinated practices. The responses of specialists to the ecological changes are the reactivity or favourable to liveliness that relies upon what sort of engineering specialist is

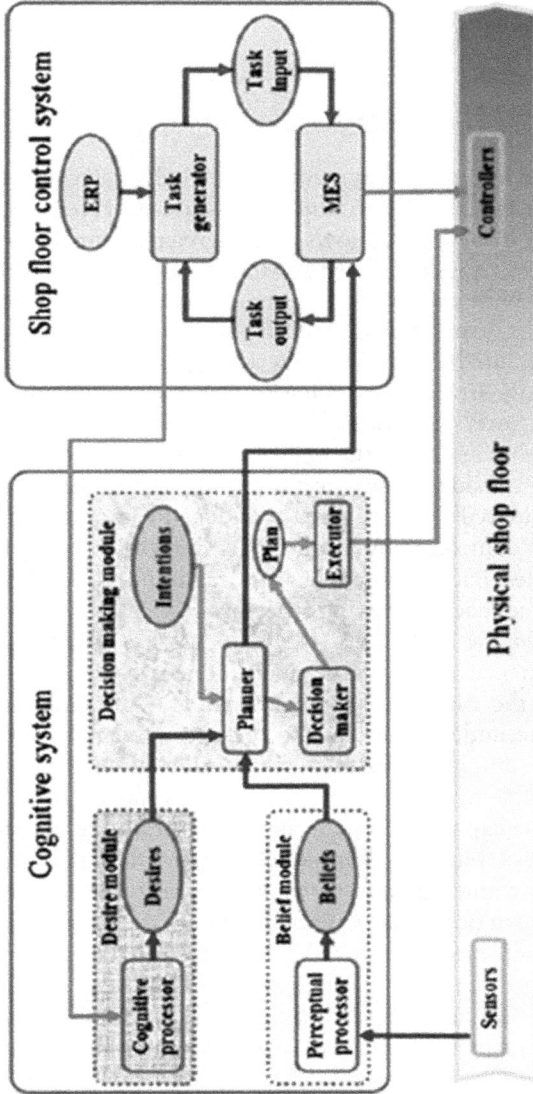

Figure 10.1 Cognitive system of manufacturing [1, 34].

utilized to foster specialists. The distinctive quality of the intellectual specialist in examination with the conventional specialist is the knowledge of the intellectual specialist, which is displayed at the improvement of the supportive of animation trademark. Insight is the capacity of the specialist utilizing its insight and thinking instruments to settle on a reasonable choice as for the ecological changes.

10.5 SWARM INTELLIGENCE

In the indigenous habitat, an aggregate knowledge is completed by basic associations of people. An idea found in the states of creepy crawlies, to be specific multitude knowledge, shows this aggregate insight. Swarm knowledge is set up from straightforward substances, which connect locally with one another and with their current circumstance. In insect provinces, the aggregate insight is given by cooperation of individual subterranean insects with the restricted intellectual capacities through synthetic substances called pheromones. To adjust with the unique advancement of climate, a multitude of subterranean insects needs the self-association capacity. Self-association is helped by revamping its design through alteration of the connections among elements without outside intercession. Moving this standard to assembling frameworks considered as a local area of independent and helpful elements, self-association is completed by locally matching between machine abilities and item necessities. Each machine has a pheromone, an incentive for a particular activity. The machine having the most elevated pheromone is picked for the activity. In the AMS-SCA, the framework self-sorts out to adjust to unsettling influences through the pheromone-based specialist exchange. In arrangement, the machine specialist dealing with the disappointment machine sends the undertaking data to the excess machine specialists. The assignment data comprises of the machining technique, the cutting conditions, and the instrument type. The machine specialists contrast this data with their machine capacity through their data set. In the information base, possible variables of a machine for completing an undertaking, for example, machine particular and capacity to machine workpiece as indicated by its useful necessities are put away. Each machine specialist is considered as an insect, and the pheromone is utilized as a correspondence middle person in specialist exchange. The capacity of pheromone is to demonstrate the capacity of machine for doing the errand generally. The pheromone esteem is utilized as the basis for picking the ideal machine among the elective machines.

10.5.1 Need for swarm intelligence

Swarm Intelligence can be described as a gathering of non-keen robots shaping, collectively, into a smart robot. All in all, a gathering of "machines" fit for shaping "requested" material examples "capriciously". Without a reasonable idea of capriciousness, the meaning of clever multitude could be applied to minor frameworks. For instance, the definition could be fulfilled by a mechanical "screen saver" that produces fascinating examples from an irregular calculation. In any case picking indiscriminately a few arranged examples from a set isn't the possibility of unconventionality that recommends insight. Consequently, whatever seems erratic essentially on the grounds that it isn't open should be precluded. In the end the contention for eccentricity runs into the computational force of the framework which is extremely fitting, since the idea of knowledge has been frequently connected with computational power, as with the Turing test, chess-playing PCs, and other AI contentions show. Now we could additionally work on the meaning of Intelligent Swarm as, Intelligent multitude: a gathering of "machines" prepared to do "unusual" material calculation.

Unconventionality can be accomplished on the off-chance that the framework making the forecast is not fit for beating the framework it is attempting to foresee. Presently, assuming that a framework is equipped for all-inclusive calculation it can't be surpassed. Indeed, if one attempts to foresee a framework which is equipped for general calculation, one should utilize one more all-inclusive machine to re-enact the first. Consequently, the limitless time conduct of a framework fit for widespread calculation is, in an everyday sense, mysterious in any limited time: the issue is officially undecidable. Typically, one does not have an approach to telling whether the strategy utilized for a specific calculation is the most effective conceivable. No reasonable lower limits on the trouble of calculation have at any point been laid out. The pace of calculation is the issue. By and large, when you have a framework that is widespread, one can make it carry out any calculation; however, the rate at which the calculation is done is not self-evident. Truth be told, without unique advancement, a general Turing machine will normally work for some proper portion of the speed of a particular Turing machine that it is set up to copy. To be sure, deduced, there can be extraordinary contrasts in the rates at which given calculations should be possible. To frame the new example the wise multitude S would have to do (1) calculations and (2) movements. The multitude does (1) and (2) simultaneously.

10.5.2 Swarm intelligence models

There are several models available for optimization. Broadly, they can be categorized upon the behavior of animals, insects, birds and amphibians. One of each of the several models under these categories is discussed below.

10.5.2.1 Butterfly optimization algorithm

Arora and Singh [35] have proposed a swarm intelligence strategy that mimics butterfly foraging and mating behaviors. This calculus is stimulated by the sensation of the smell of butterflies. The sense of smell helps butterflies reveal their accomplices and their food. Soon, target work with the underlying population was introduced. Here, the mass remains fixed and the memory is distributed. After that, areas of butterflies are randomly created, and their scent and wellness values are registered and stored. Focusing work involves repositioning numerous butterflies in the design space and determining their well-being. The final placement is related to the ideal wellness assessment. In a global pursuit, butterflies capture the scents of other butterflies and chase them forward. Conversely, when hunting nearby, butterflies cannot sense the scent and move spontaneously. The power of aroma plays an important role in both local and global research. World Hunt plans to track the most appropriate placement of butterflies. Butterfly Optimization Algorithm has been observed to be easy to use and implement. It is unaware of many borders and can quickly merge into the ideals of the world without premature mergers. In addition, it is very helpful in finding the ideal answer to a noisy problem. In addition, it is extremely powerful in solving problems with multiple unbiased or global improvements.

10.5.2.2 Harris Hawk optimization

A population-based procedure to copy the hunting behaviour of the Harris Hawk is presented in [17]. The main motivation is the collaboration between the Harris Hawk and his hunting style (shock jump). They attack their prey at the rally. They have two hunting techniques for attacking their prey. The tricky blocking method runs out of prey and eventually hits the prey. In hard blocks, when the prey is exhausted, the Harris Peddle traps it for an unexpected attack. These attack strategies model the exploit phase, and prey tracking is the investigative phase. Harris hawk optimization's multi-step planning in the abuse phase allows for short, random examples of persecution. Therefore, if one attack method does not work, you can use another method to save the best method for the next cycle. It is incredibly useful in the issue of improvement as it guarantees a steady harmony between investigation (global pursuit) and abuse (neighbourhood investigation). Depending on the starting energy of the prey, Harris Hawk optimization may guide the action from investigation to double-play and then change the double-play activity. Harris Hawk optimization uses versatile and time-varying boundaries to monitor search range issues such as near-optimal, multiple remoteness, and operational optimality. In the case of multivariable questions, this calculation helps to adjust for exploratory and dubious examples. For high-level issues, it provides a faster and more powerful deployment than other proven improvements approaches. Harris Hawk optimization uses time-shifted parts to make investigations and ambiguities in reruns more stable. This depends on improving the slope less population. This gives you an advantage over other methods in terms of conversion speed.

10.5.2.3 Red deer algorithm

The Red deer algorithm created by Fathollahi-Fard et al. [36] depends on thundering and special mating conduct. While viewing for a mating a potential open door, male Red deer produce a "thunder" or "wail" and charge each other to draw in the group of concubines with additional hinds. A collection of mistresses is a local area of females that is overwhelmed by the male leader. Male deer are separated into bunches called authority and stags in view of their thundering limit. This seriousness between male red deer to acquire prevalence over females is the primary key of the calculation and is accomplished in two stages—intensification and diversification. Double-dealing (intensification) is accomplished w.r.t thundering of male deer and fighting among authority and stage. In the arrangement space, the place of the male red deer's neighbours is determined and, assuming their goal capacities are better, it gets refreshed. Investigation is accomplished w.r.t development of collections of mistresses and dispensing to the commandant according to their power. Alongside it, the mating of collection of mistress commandant with a level of hinds in his group of concubines and another collection of mistresses works on the investigation.

10.5.2.4 Whale optimization algorithm

Mirjalili and Lewis [17] fostered another shallow water-savvy technique that reproduces the communism of humpback whales while chasing after prey. Rather than other metaheuristics, whale streamlining calculations imitate whale hunting systems and unique assault attributes (bubble nets). As whales circle their prey, certain air pockets structure in a roundabout way. Bubble net assault innovation is helpful for double-dealing. The prey search condition of the whale improvement calculation addresses the inquiry stage. Because of double-dealing, the whale's position is refreshed with either a winding development or a contracting attack.

10.6 MANUFACTURING SYSTEM MODELLING

Several researchers [37–47] worked on the modeling of manufacturing systems. A multitude of intellectual specialists is to be created to keep the assembling system running when aggravations show up. Assets of the machining shop are constrained by specialists on account of aggravations. Every specialist defeats the aggravation without help from anyone else or helps different specialists through remote correspondence to beat the unsettling influence. In any case, the manufacturing execution system controls the manufacturing site. Specialists are separated into practical intellectual specialists such as toolmaking specialists, career specialists, machine specialists, and robot specialists. This department is based on the capabilities undertaken by the experts in the machining plant. MES data for executive booking and organization, and dispatch and execution, are used to plan interactions and activities, and to review cycles and schedules. Machine specialist data includes data on the physical and interactive capabilities of your machine, as well as thought-independent steering data. This data is used to complete the processing process performed and to collaborate with various professionals. The data of workpiece, carrier, and robot specialist conveys their own information to help the choices of different specialists and impart for choosing where to go [34, 48–54].

10.7 THE OPERATION OF AUTONOMOUS MANUFACTURING SYSTEM IN LIGHT OF SWARM OF COGNITIVE AGENTS (AMS-SCA)

Toward the start, the MES sends an assignment order to both the regulators and the intellectual specialist. The intellectual processor recognizes the objectives and transforms to the cravings. The state of the processing hole is updated by the test module. This module then, at that point, channels information to get the data comparing to liabilities of the specialist. On the off-chance that unsettling influences happen, the information is categorized into high and low frequencies by the component removal unit. Determination results report conditions of the machining shop: aggravations or typical status. The organizer looks at the information from the result of the determination module with the ideal objectives. Assuming that the information arrives at the ideal objectives, the MES sends a message reporting the health of the machine and the workshop continues to run. In any case the chief creates another arrangement dependent on the information, wants and expectations. This arrangement is straightforwardly finished by the machine wherein the aggravation happens when the unsettling influence is not difficult to recuperate. In the event that the unsettling influence is hard to recuperate or needs administrator intercession, for example, the machine breakdown, this arrangement is finished by another machine. The intellectual specialist executes an exchange with different specialists. It sends a solicitation for help to all machine specialists. The best arrangement is picked dependent on the assessment of pheromone upsides of the elective machine specialists in the event that many machine specialists fulfil the prerequisites. Crafted by the machine wherein the unsettling influence happens is performed at one more machine to keep the assembling framework running. The chose specialist makes an impression on the tool specialist and the carrier specialist to report that it plays out crafted by the machine where the unsettling influence occurs. The machining framework utilizes the past arrangement when the upset machine is re-established. This arrangement is applied for unsettling influences, which need a short recuperating time. On the off-chance that the aggravations need a long recuperating time or the arrangement between specialists doesn't have any arrangement, the solicitation is shipped off the MES for rescheduling.

10.8 THE IMPLEMENTATION OF AUTONOMOUS MANUFACTURING SYSTEM IN LIGHT OF SWARM OF COGNITIVE AGENTS (AMS-SCA)

There are three gatherings of aggravations: rescheduling, non-arrangement, and exchange bunch. In the thought of going to lengths, the rescheduling bunch implies that the doled-out machining undertaking ought to be rescheduled because of the long recuperation time, for example over one hour while halting the entire framework. This given time depends on the impact of unsettling influence to the arranged timetable of the considered machining framework. Throughout this time, it is exceptionally difficult to keep the arranged timetable inside the restricted resistance. The non-exchange bunch has a place with the aggravations of which the recuperating time is under 30 minutes and the strategies for recuperation are known from the past experience. The given time for characterizing non-exchange or arrangement bunches depends on the measurements of aggravations when machining grip lodgings. Nearly unsettling influences for which the recuperating time is under 30 minutes are not difficult to be recuperated, and can be fixed by an administrator with his own insight. The rest of unsettling influences is gathered to the exchange issue. Those unsettling influences can be addressed with the information gathered while working the ordinary machining framework through the specialist arrangement process.

The instruments of the Autonomous Manufacturing System in light of Swarm of Cognitive Agents for adjusting to aggravations, which have a place with non-exchange and arrangement bunches are demonstrated by executing the AMS-SCA [1] proving ground. In a proving ground every one of the Personal Computers (PCs) incorporates three specialists, for example, the machine, work-piece, and carrier specialist, which are liable for dealing with the machining system of a work-piece utilizing a machine apparatus. The work-piece specialist gets the data of the work-piece from a RFID tag through a RFID peruser. The carrier specialist gets a handle on the present status to work a carrier from the sensors. To save the speculation and establishment endeavors, unsettling influence generators (turn on/off switches) are utilized to produce aggravations. The equipment parts like Programmable Logic Controllers (PLCs), RFID perusers are associated with the specialist by wire strategy. The PLCs considered as regulators of machine instruments get the interaction data from the MES framework and execute the machining position. Remote correspondence is utilized to make a connection between intellectual specialists, and machine specialists with the MES framework. The communication is finished by handling messages. The eXtensible Markup Language (XML) is utilized to arrange the directives for deciphering the information. The arrangement of specialists is carried out by utilizing the subterranean insect settlement strategy.

The framework design of the intellectual specialist-based machining shop directs out the three piece issues toward execute the intellectual specialists, which are the cooperation convention, specialist practices, and data set (DB) just as the data stream among parts in the framework for completing the usefulness of the proposed AMS-SCA. The specialist connects with the MES and different specialists through the XML messages. The cycle control convention (OPC) for connecting and implanting objects is utilized for discussing the specialist with PLC which associate with the actual gadgets on the machining shop, for example, sensors, unsettling influence information, and caution gadget. The data sets, including the cycle data, the specialist addresses for conveying in the organization, the pheromone upsides of the assignments connected with the machine specialists. The specialist utilizes the "search" strategy to analyze and characterize the aggravation. With regard to the unsettling influence type, the specialist motivations to settle on a choice utilizing the "change" or "coordinated effort" strategies. In cooperation, the specialists produce the pheromone worth of the doled-out task utilizing the "work out" technique. Then, at that point, the "arrange" process is completed among specialists to observe the specialist with the most noteworthy pheromone an incentive for doing the assignment.

The machining framework comprising of three machining focuses is considered to apply to the proving ground. In the event that the unsettling influence occurs on one machine, which has a place with the non-exchange bunch, the comparing intellectual specialist utilizes the self-change instrument to conquer the aggravation. In any case, the unsettling influence has a place with the arrangement bunch, the specialist haggles with different specialists utilizing the pheromone-based exchange system.

10.9 PRODUCTION PLANNING AND CONTROL

As an essential for presenting the item as an extra component of creation control, solid admittance to applicable item explicit data, reliable task and the standard update of the item's state should be guaranteed. The use of RFID innovation gives the premise to an exact picture of this present reality and supports independent decision-production in disseminated and computerized creation control. RFID shows a few benefits contrasted with other mechanized recognizable proof innovations, for example, programmed and remote machine-to-machine correspondence, or the dynamic and programmed putting away and recovering of data. In this manner, things can be followed consequently and basic item explicit data can be put away straightforwardly on the item. By joining tools or items with RFID transponders, they become "shrewd items." They give the resources to impart their data to the arranging level and with different assets through an individual correspondence foundation. Henceforth, brilliant items can naturally handle obtained demands and react to the accessible RFID peruses. Shrewd items are consequently ready to effectively impact their own creation, circulation and capacity. The correspondence between the shrewd item and assets throughout the creation cycle as well as the item's incorporation in the general interaction is portrayed in Figure 10.2

It incorporates the persistent observing of brilliant items and assets (e.g., state data, functional accessibility) and the particular exchange of the refreshed data from the shop floor to the arranging level. Item-focused information from the board is a vital essential for a decentralized item-based control of creation processes. It comprises of the separate meaning of pivotal item explicit data that should be put away on the savvy item. This data should be legitimately organized on the RFID transponder, to guarantee dependable and normalized processes as well as a simple trade and adjustment of item-explicit information. A normalized XML outline was created for the portrayal of item-explicit information, including ID information, handling information and quality information. The single specialists in a MAS are unnecessarily subject to refreshed ongoing data about assets also

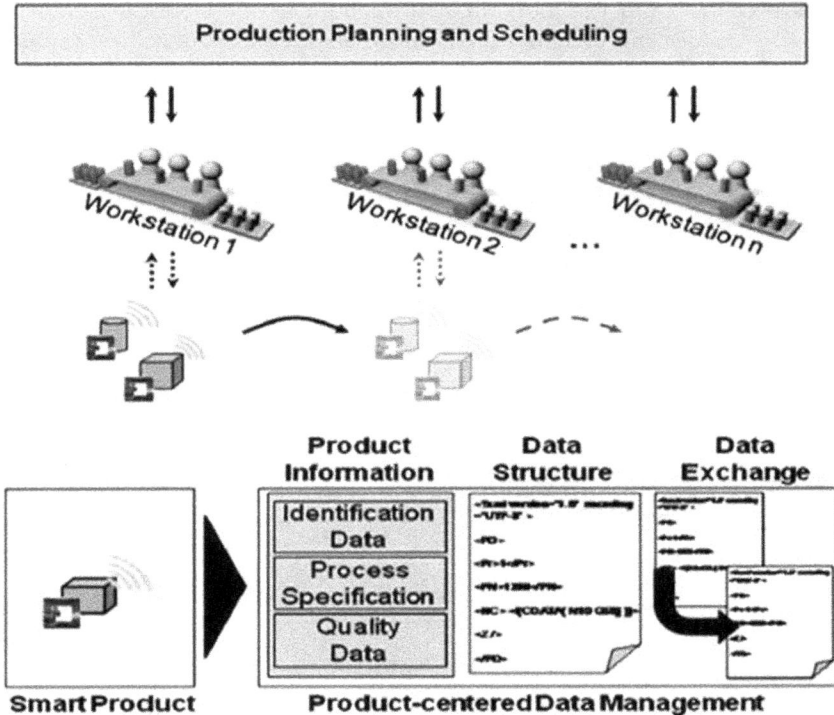

Figure 10.2 Production planning and control [22].

items from the shopfloor, in this manner giving an ideal execution of creation arranging and control. In particular, the necessities that arise are related to item customization, recognizability, and the data board. Subsequently, the perceptibility, quality, accessibility, and usefulness of the creation framework can be upgraded with further developed productivity of reaction to unsettling influences in process control. By utilizing the previously mentioned savvy item, a pertinent constant data trade between an item in the creative climate and its virtual partner in a multi-agent system (MAS) can be understood. Each expert workpiece is furnished with an RFID transponder, on which basic item-explicit assembling data, for example, the work number, the underlying work plan, quality information, and the creation history is put away.

Figure 10.3 shows a commendable multi-specialist upheld, item-based creation process. The proposed MAS comprises of arranging specialists, item specialists and machine specialists for the preparation and control of creation processes. In light of approaching creation arrangements, the arranging specialist delivers each request for the creation cycle and moves the singular request to the item specialist. Past the gig discharge, the arranging specialist should additionally consider possibly adjusted limit conditions. The item specialist is accountable for the ensuing creation of the workpiece/item and the individual throughput in the creation climate. It should consider the predetermined work plan, confirm machine capacity and arrange the handling with the machine specialists. Furthermore, the item specialist, which is simultaneously a quality specialist, needs to assess possible deviations among arranged and executed cycles and the separate workpiece quality. The machine specialist haggles with the item/quality specialist and balances the solicitation for handling with the accessible asset limits. In addition, the machining grouping is upgraded with the new expansion of a creation request.

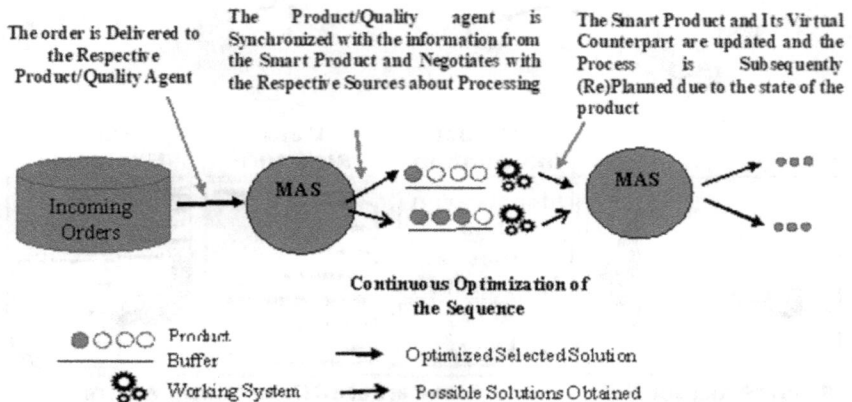

Figure 10.3 Multi agent system (MAS) supported production [22].

Decentralized item-based information the board considers a direct and blunter safe collaboration between mechanized creation processes and possible manual intercessions of human specialists, for example, the modification of workpieces. The total documentation of executed cycle steps and quality data with the constant update takes into account the immediate thought in resulting creation processes and practical adjustments. Accordingly, essential creation cycles can be resolved consequently founded on the assigned work process and its present status in the assembling climate. The singular specialists are in this manner ready to get to, oversee and use the data carried on the RFID transponders and apply it to creation arranging and control, notwithstanding the practical manual intercessions of human laborers. On account of manual undertakings, for example, gathering processes or adjust, the workpiece can be taken out from the computerized creation process. The human specialist can involve item-explicit data as work directions. In the wake of completing the manual work steps, the administrator can report his work on the brilliant item and return it to the robotized creation cycle. With normalization, item-focused information the executives, the brilliant item can be consequently reintegrated into the control of the MAS. In a similar methodology, it is feasible to physically dispatch a creation request to the framework. In this way, uncommon degrees of adaptability can be achieved underway conditions with a mix of mechanized and manual cycles.

10.10 THE NEED FOR MULTI-AGENT SYSTEMS (MAS)

In the Distributed Manufacturing System (DMS) hierarchical model a few fundamental highlights between elements are available at all times: independence, collaboration, adaptability, flexibility, insight, and so on. Independent specialists cooperating can make evolvable frameworks. A MAS is fundamentally a framework formed by a larger number of people and different independent specialists that together have the limit of arriving at the general objectives of the framework where they are embedded. These specialists can have two particular methodologies:

- Practical disintegration: specialists are utilized to typify modules point by point to capacities (request securing, process arranging and booking, material dealing with).
- Actual disintegration: specialists are utilized to address the actual world (laborers, machines, devices, tasks).

The useful methodology embroils the divide between independent specialists of numerous factors among numerous and various capacities that can initiate a few irregularities. Then again, in the actual methodology, the specialists have fewer factors to share and in this way more effortlessness in his singular administrations. In any case, the huge number of specialists that it is fundamental, can convey new issues like correspondence upward and a complicated specialist the board is need. A MAS offers many benefits in a DMS execution; to be practicable in terms of execution, a decent design for the MAS association and specialist epitome is required, and it is important a right decision of the conventions for correspondence, participation and arrangement.

In MAS scheduling, specialists control asset or potentially request factors under their own position. Potential work communications are taken care of by specialists letting each know other their total responsibilities and free limits. Since worldwide consistency can scarcely be ensured, requirement checking or re-enactment is required. Since limitation unwinding and back-tracking may require tremendous computational overheads, their utilization is restricted. As an option in contrast to backtracking, heuristic fiddling of nearly struggle free timetables can be applied.

10.11 CONCLUSION

The multitude of intellectual specialists builds the vigor of the framework by keeping away from incorporated control and showing the capability of carrying out independent practices by adaptable capacity in choice making. The assembling control framework, dependent on intellectual specialists, has sufficient capacity to adjust independently to unsettling influences without upper-level guides or an all-out alteration of the arrangements. Then again, the assembling control framework furnished with the fake intellectual capacities meets the necessities of adaptability, flexibility and unwavering quality. The advancement of an Autonomous Manufacturing System dependent on Swarm of Cognitive Agents (AMS-SCA) to adjust to inner unsettling influences in an independent manner is the new commitment of this exploration. This innovation empowers the relevance of intellectual practices of humans to defeat the unsettling influences inside the machining framework. To demonstrate the proficiency of the proposed AMS-SCA idea, the proving ground was carried out and zeroed in on the self-change instrument on account of the unsettling influences, which have a place with the non-exchange bunch, just as the pheromone-based system for adjusting to aggravations, which have a place with the arrangement bunch.

REFERENCES

[1] H-S Park, N-H Tran. An autonomous manufacturing system based on swarm of cognitive agents. *Journal of Manufacturing Systems* (2012). http://dx.doi.org/10.1016/j.jmsy.2012.05.002
[2] AR Cheraghi, S Shahzad, K Graffi. Past, present, and future of swarm robotics. arXiv:2101.00671v1 [cs.RO] 3 Jan 2021.

[3] ZH Ismail, MGM Hamami. Systematic literature review of swarm robotics strategies applied to target search problem with environment constraints. *Applied Sciences* (2021). https://doi.org/10.3390/app11052383

[4] I Navarro, F Matía. An introduction to swarm robotics. *International Scholarly Research Notices* (2012). http://dx.doi.org/10.5402/2013/608164

[5] G Beni. From swarm intelligence to swarm robotics. *Swarm Robotics WS 2004*, LNCS 3342. 2005. pp. 1–9.

[6] Y Cai, A Sharma. Swarm intelligence optimization: An exploration and application of machine learning technology. *Journal of Intelligent Systems* (2020). https://doi.org/10.1515/jisys-2020-0084

[7] J Guo, M Martínez-García. Key technologies towards smart manufacturing based on swarm intelligence and edge computing. *Computers & Electrical Engineering* (2021). https://doi.org/10.1016/j.compeleceng.2021.107119

[8] P Monga, M Sharma, SK Sharma. A comprehensive meta- analysis of emerging swarm intelligent computing techniques and their research trend. *Journal of King Saud University - Computer and Information Sciences* (2021). https://doi.org/10.1016/j.jksuci.2021.11.016

[9] M Schranz, M Umlauft, M Sendel, W Elmenreich. Swarm robotic behaviors and current applications. *Frontiers in Robotics and AI* (2020). https://doi.org/10.3389/frobt.2020.00036

[10] K Kaur, Y Kumar. Swarm intelligence and its applications towards various computing: A systematic review. *2020 International Conference on Intelligent Engineering and Management (ICIEM)*. ©2020 IEEE. (2020). pp. 57–62.

[11] G Ramanana, DBP Rajab, K Samuel. Intelligent particle swarm optimization for quality machining in advanced manufacturing process. *Materials Today: Proceedings* 24 (2020) 510–518.

[12] TV Sibalija. Particle swarm optimization in designing parameters of manufacturing processes: A review (2008–2018). *Applied Soft Computing* (2019). https://doi.org/10.1016/j.asoc.2019.105743

[13] J Qin, Y Liu, R Grosvenor, F Lacan, Z Jiang. Deep learning- driven particle swarm optimization for additive manufacturing energy optimization. (2019). https://doi.org/10.1016/j.jclepro.2019.118702

[14] T Navalertporn, NV Afzulpurkar. Optimization of tile manufacturing process using particle swarm optimization. *Swarm and Evolutionary Computation* (2011). https://doi.org/10.1016/j.swevo.2011.05.003

[15] Q Wu. Product demand forecasts using wavelet kernel support vector machine and particle swarm optimization in manufacture system. *Journal of Computational and Applied Mathematics* (2010). https://doi.org/10.1016/j.cam.2009.10.030

[16] O Durán, N Rodriguez, LA Consalter. Collaborative particle swarm optimization with a data mining technique for manufacturing cell design. *Expert Systems with Applications* (2010). https://doi.org/10.1016/j.eswa.2009.06.061

[17] S Mirjalili, A Lewis. The whale optimization algorithm. *Advances in Engineering Software* 95 (2016): 51–67.

[18] G Guerreiro, R Costa, P Figueiras, D Graça, RJ Gonçalves. A self-adapted swarm architecture to handle big data analytics for factories of the future. *IFAC-PapersOnLine* (2019). https://doi.org/10.1016/j.ifacol.2019.11.356

[19] R Wang, Y Feng. Evaluation research on green degree of equipment manufacturing industry based on improved particle swarm optimization algorithm. *Chaos, Solitons & Fractals* (2019). https://doi.org/10.1016/j.chaos.2019.109502

[20] M Chamanbaz, D Mateo, BM Zoss, G Tokic, E Wilhelm, R Bouffanais, DKP Yue. Swarm-enabling technology for multi-robot systems. *Frontiers in Robotics and AI* (2017). https://doi.org/10.3389/frobt.2017.00012

[21] G Dominic. The holonic approach for flexible production a theoretical framework. *Elixir Production* 42 (2012) 6106–6110.

[22] A Bannat, T Bautze, M Beetz, et al. Artificial cognition in production systems. *IEEE Transactions on Automation Science and Engineering* 8 (2010) 148–174.

[23] K Shea, C Ertelt, T Gmeiner, F Ameri. Design-to-fabrication automation for the cognitive machine shop. *Advanced Engineering Informatics* 24 (2010) 251–268.

[24] PP Leitao. Agent-based distributed manufacturing control: A state-of-the-art survey. *Engineering Application of Artificial Intelligence* 22 (2009) 979–991.

[25] YF Wang, YF Zhang, JYH Fuh, ZD Zhou, P Lou, LG Xue. An integrated approach to reactive scheduling subject to machine breakdown. *Proceeding of the IEEE International Conference on Automation and Logistics.* 2008. pp. 542–547.

[26] D Anghinolfi, A Boccalatte, A Grosso, M Paolucci, A Passadore, C Vecchiola. A swarm intelligence method applied to manufacturing scheduling (2007). https://www.researchgate.net/publication/220866173

[27] HV Brussel, L Bongaerts, J Wyns, P Valckenaers, TV Ginderachter. A conceptual framework for holonic manufacturing: Identification of manufacturing holons. *Journal of Manufacturing Systems* 18 (1999) 35–52.

[28] BH Zhou, CC Li, X Zhao. FIPA agent-based control system design for FMS. *International Journal of Advanced Manufacturing Technology* 31 (2007) 969–997.

[29] L Monostori, J Váncza, SRT Kumara. Agent-based systems for manufacturing. *Annals of the CIRP* 55 (2006) 697–720.

[30] E Westkampfer. Manufacturing on demand in production networks. *Annals of the CIRP* 46 (1997) 329–334.

[31] GE Vieira, JW Hermann, E Lin. Rescheduling manufacturing systems: A framework of strategies, policies, and methods. *Journal of Scheduling* 6 (2003) 39–62.

[32] S Arora, S Singh. Butterfly optimization algorithm: A novel approach for global optimization. *Software Computing* 23(3) (2019) 715–734.

[33] L Monostori. AI and machine learning techniques for managing complexity, changes and uncertainties in manufacturing. *Engineering Applications of Artificial Intelligence* 16 (2003) 277–291.

[34] FS Nobre, AM Tobias, DS Walker. The pursuit of cognition in manufacturing organizations. *Journal of Manufacturing Systems* 27 (2008) 145–157.

[35] AA Heidari, S Mirjalili, H Faris, I Aljarah, M Mafarja, H Chen. Harris hawks optimization: Algorithm and applications. *Future Generation Computer Systems* 97 (2019) 849–872.

[36] AM Fathollahi-Fard, M Hajiaghaei-Keshteli, R Tavakkoli-Moghaddam. Red deer algorithm (RDA): A new nature-inspired meta-heuristic. *Soft Computing* 24 (2020) 14637–14665. https://doi.org/10.1007/s00500-020-04812-z. P Valckenaers, H. Van Brussel. Holonic manufacturing execution systems. *Annals of the CIRP* 54 (2005) 427–432.

[37] M Saadat, MCL Tan, M Owliya. Changes and disturbances in manufacturing systems: A comparison of emerging concepts. *World Autom Congress Proceedings.* 2008. pp. 556–560.

[38] HS Park, HW Choi. Development of a modular structure-based changeable manufacturing system with high adaptability. *International Journal of Precision Engineering and Manufacturing* 9 (2008) 7–12.

[39] P Leitao, F Restivo. ADACOR: A holonic architecture for agile and adaptive manufacturing control. *Computers in Industry* 57 (2006) 121–130.

[40] A Tharumarajah, AJ Wells, L Nemes. Comparison of emerging manufacturing concepts. *IEEE International Conference on Systems, Man and Cybernetics.* 1998. pp. 325–331.

[41] C Christo, C Cardeira. Trends in intelligent manufacturing systems. *Proceedings of the IEEE International Symposium on Industrial Electronics.* 2007. pp. 3209–3214.

[42] L Monostori, B Kadar, J Hornyak. Approaches to managing changes and uncertainties in manufacturing. *Annals of the CIRP* 47 (1998) 365–368.

[43] G Langer, L Alting. An architecture for agile shop floor control systems. *Journal of Manufacturing Systems* 19 (2000) 267–281.

[44] MF Zaeh, M Beetz, K Shea, et al. The cognitive factory. In ElMaraghy HA, editor. *Changeable and Reconfigurable Manufacturing Systems.* 2009. pp. 355–371, Springer.

[45] B Denkena, H Henning, LE Lorenzen. Genetics intelligence: New approaches in production engineering. *Production Engineering Research and Development* 4 (2010) 65–73.

[46] MF Zaeh, C Lau, M Wiesbeck, M Ostgathe, W Vogl. Towards the cognitive factory. *Proc Int Conf CARV.* 2007.

[47] X Zhao, Y Son. BDI-based human decision-making model in automated manufacturing systems. *International Journal of Modelling and Simulation* 28 (2008) 347–356.

[48] K Ueda, T Kito, N Fujii. Modeling biological manufacturing system with bounded- rational agents. *Annals of the CIRP* 55 (2006) 469–472.

[49] K Ueda. *Emergent Synthesis Approaches to Biological Manufacturing Systems.* 2007. Digital Enterprise Technology.

[50] K Ueda, I Hatono, N Fujii, J Vaario. Reinforcement learning approaches to biological manufacturing systems. *Annals of the CIRP* 49 (2000) 343–346.

[51] HP Wiendahl, HA ElMaraghy, P Nyhuis, MF Zaeh, HH Wiendahl, N Duffie, M Brieke Changeable manufacturing-classification, design and operation. *Annals of the CIRP* 56 (2007) 783–809.

[52] HAEI Maraghy. Flexible and reconfigurable manufacturing systems paradigms. *International Journal of Flexible Manufacturing Systems* 17 (2006) 261–271.

[53] T Tolio, D Ceglarek, HA ElMaraghy, et al. SPECIES-Co-evolution of products, processes and production systems. *Annals of the CIRP* 59 (2010) 672–693.

[54] P Leitao, F Restivo. Agent-based holonic production control. *Proc Int Workshop Database Expert Syst Appl.* 2002. pp. 589–596.

Index

For Product Safety Concerns and Information please contact our EU
representative GPSR@taylorandfrancis.com
Taylor & Francis Verlag GmbH, Kaufingerstraße 24, 80331 München, Germany